AF598023

# METHODS IN MOLECULAR BIOLOGY



For further volumes:
http://www.springer.com/series/7651

# Plant Epigenetics and Epigenomics

## Methods and Protocols

Edited by

**Charles Spillane**

*Genetics & Biotechnology Lab, Plant & Agribiosciences Centre (PABC), School of Natural Sciences, National University of Ireland, Galway (NUI Galway), Ireland*

**Peter C. McKeown**

*Genetics & Biotechnology Lab, Plant & Agribiosciences Centre (PABC), School of Natural Sciences, National University of Ireland, Galway (NUI Galway), Ireland*

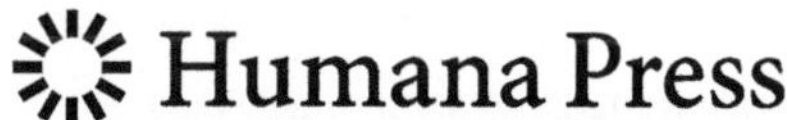

*Editors*
Charles Spillane
Genetics & Biotechnology Lab
Plant & Agribiosciences Centre (PABC)
School of Natural Sciences
National University of Ireland
Galway (NUI Galway), Ireland

Peter C. McKeown
Genetics & Biotechnology Lab
Plant & Agribiosciences Centre (PABC)
School of Natural Sciences
National University of Ireland
Galway (NUI Galway), Ireland

ISSN 1064-3745 ISSN 1940-6029 (electronic)
ISBN 978-1-62703-772-3 ISBN 978-1-62703-773-0 (eBook)
DOI 10.1007/978-1-62703-773-0
Springer New York Heidelberg Dordrecht London

Library of Congress Control Number: 2013958314



Printed on acid-free paper

Humana Press is a brand of Springer
Springer is part of Springer Science+Business Media (www.springer.com)

# Preface

"Treasure your exceptions! When there are none, the work gets so dull that no one cares to carry it further. Keep them always uncovered and in sight. Exceptions are like the rough brickwork of a growing building which tells that there is more to come and shows where the next construction is to be." Geneticist William Bateson offered this advice in 1908, around the dawn of modern genetics following the rediscovery of Gregor Mendel's pea plant experiments, and it remains sound today.

Modern molecular biologists have access to the complete genome sequences of many species of interest, including many species of crops and other plants. To fully understand the natural history of an organism and its potential for change under natural selection requires understanding of how these genomes are regulated during growth, differentiation, and reproduction. It is now appreciated that these processes are affected in key ways by the epigenome which orchestrates genomic organization, expression and repair, and interacts with networks of gene, protein and metabolite regulation during eukaryote development. Many of the fundamental discoveries concerning the mechanisms of epigenetic regulation have arisen from studies performed in plants, often due to the investigation of phenomena which had initially been regarded merely as curiosities, the general relevance of which only later became clear. Discoveries made in this way range from transposons and nucleolar dominance to paramutation, and the inducible gene silencing which led to the discovery of RNAi.

This volume of "Methods in Molecular Biology" gathers together comprehensive descriptions of the techniques currently being used to define the details of the plant epigenetic landscape. Such a work is timely, as the number of sequenced plant genomes is rapidly increasing. The activity of these genomes is controlled by covalent modification, packaging with histones and chromatin-remodelling proteins, and the activity of small RNAs which together define the epigenome. We have concentrated especially upon the application of recently developed techniques to analyze plant phenomena with known epigenetic components, such as flowering time, imprinting, and dosage effects. We have drawn upon the expertise of colleagues applying contemporary high-throughput screens, microscopy, and bioinformatic techniques to laboratory models, notably *Arabidopsis thaliana*, although the techniques presented are applicable for studies in crops and non-model species of evolutionary or ecological significance.

It is our hope that these reviews of contemporary methods will advance the study of plant epigenetic phenomena, and allow the biological community to fully integrate our understanding of epigenetic mechanisms into models of plant function during development and evolution.

***Galway, Ireland***

***Charles Spillane***
***Peter C. McKeown***

# Contents

## Contributors

MALIKA AINOUCHE • *Université Rennes 1, Rennes, France*
EMIDIO ALBERTINI • *Department of Applied Biology, University of Perugia, Perugia, Italy*
JAMES A. BIRCHLER • *Division of Biological Sciences, University of Missouri, Columbia, MO, USA*
Z. JEFFREY CHEN • *Section of Molecular Cell Developmental Biology, Center for Computational Biology and Bioinformatics, Austin, TX, USA; Institute for Cellular and Molecular Biology, The University of Texas at Austin, Austin, TX, USA*
MARIA COLOMÉ-TATCHÉ • *Faculty of Mathematics and Natural Sciences, Groningen Bioinformatics Centre, University of Groningen, Groningen, The Netherlands*
VINCENT COLOT • *Institut de Biologie de l'Ecole Normale Supérieure, Centre National de la Recherche Scientifique (CNRS), Institut National de la Santé et de la Recherche Médicale (INSERM), Paris, France*
SANDRA CORTIJO • *Institut de Biologie de l'Ecole Normale Supérieure, Centre National de la Recherche Scientifique (CNRS), Institut National de la Santé et de la Recherche Médicale (INSERM), Paris, France*
ROBERT DAY • *Department of Biochemistry, University of Otago, Dunedin, New Zealand*
CAROLINE DEAN • *Cell & Developmental Biology, John Innes Centre, Norwich, Norfolk, UK*
CHRISTOPHER DEFRAIA • *Department of Molecular Genetics, The Ohio State University, Columbus, OH, USA*
ANN DEPICKER • *Department of Plant Systems Biology, VIB, Ghent University, Ghent, Belgium; Department of Plant Biotechnology and Genetics, Ghent University, Ghent, Belgium*
ANTOINE FORT • *Genetics & Biotechnology Lab, Plant & Agribiosciences Centre (PABC), School of Natural Sciences, National University of Ireland, Galway (NUI Galway), Ireland*
MARIE-ANGÈLE GRANDBASTIEN • *Institut Jean-Pierre Bourgin, INRA Centre de Versailles-Grignon, Versailles, France*
MATHIEU INGOUFF • *Faculté des Sciences, Université Montpellier2, Montpellier, France*
FRANK JOHANNES • *Faculty of Mathematics and Natural Sciences, Groningen Bioinformatics Centre, University of Groningen, Groningen, The Netherlands*
CLAUDIA KÖHLER • *Department of Plant Biology and Forest Genetics, Uppsala BioCenter, Swedish University of Agricultural Sciences, Uppsala, Sweden*
PAUL K. KORIR • *School of Mathematics, Statistics and Applied Mathematics, National University of Ireland, Galway (NUI Galway), Ireland*
ERIC LAM • *Department of Plant Biology & Pathology, Rutgers the State University of New Jersey, New Brunswick, NJ, USA*
CHONGYUAN LUO • *Department of Plant Biology & Pathology, Rutgers the State University of New Jersey, New Brunswick, NJ, USA*
RICHARD MACKNIGHT • *Department of Biochemistry, University of Otago, Dunedin, New Zealand*

GIANPIERO MARCONI • *Department of Applied Biology, University of Perugia, Perugia, Italy*
PETER C. MCKEOWN • *Genetics & Biotechnology Lab, Plant & Agribiosciences Centre (PABC), School of Natural Sciences, National University of Ireland, Galway (NUI Galway), Ireland*
GYOUNGJU NAH • *Section of Molecular Cell and Developmental Biology and Center for Computational Biology and Bioinformatics, The University of Texas at Austin, Austin, TX, USA*
DANNY W-K. NG • *Section of Molecular Cell and Developmental Biology and Center for Computational Biology and Bioinformatics, The University of Texas at Austin, Austin, TX, USA*
CHRISTIAN PARISOD • *Laboratory of Evolutionary Botany, Biology Institute, University of Neuchâtel, Neuchâtel, Switzerland*
BAS RUTJENS • *Molecular Genetics Group, University of Utrecht, Utrecht, The Netherlands*
ARMEL SALMON • *Université Rennes 1, Rennes, France*
CATHAL SEOIGHE • *School of Mathematics, Statistics and Applied Mathematics, National University of Ireland, Galway (NUI Galway), Ireland*
XIAOLI SHI • *Section of Molecular Cell and Developmental Biology and Center for Computational Biology and Bioinformatics, The University of Texas at Austin, Austin, TX, USA*
R. KEITH SLOTKIN • *Department of Molecular Genetics, The Ohio State University, Columbus, OH, USA*
JIE SONG • *Cell & Developmental Biology, John Innes Centre, Imperial College London, London SW7 2AZ*
CHARLES SPILLANE • *Genetics & Biotechnology Lab, Plant & Agribiosciences Centre (PABC), School of Natural Sciences, National University of Ireland, Galway (NUI Galway), Ireland*
REINER A. VEITIA • *Institut Jacques Monod, CNRS and Universite Paris-Diderot, Paris, France*
LEEN VERMEERSCH • *Department of Plant Systems Biology, VIB, Ghent University, Ghent, Belgium; Department of Plant Biotechnology and Genetics, Ghent University, Ghent, Belgium*
RENÉ WARDENAAR • *Faculty of Mathematics and Natural Sciences, Groningen Bioinformatics Centre, University of Groningen, Groningen, The Netherlands*
ISABELLE WEINHOFER • *Department of Biology and Zurich-Basel Plant Science Center, Swiss Federal Institute of Technology, ETH Centre, Zurich, Switzerland*
NANCY DE WINNE • *Department of Plant Systems Biology, VIB, Ghent University, Ghent, Belgium; Department of Plant Biotechnology and Genetics, Ghent University, Ghent, Belgium*

# Chapter 1

# Landscaping Plant Epigenetics

## Peter C. McKeown and Charles Spillane

## Abstract

The understanding of epigenetic mechanisms is necessary for assessing the potential impacts of epigenetics on plant growth, development and reproduction, and ultimately for the response of these factors to evolutionary pressures and crop breeding programs. This volume highlights the latest in laboratory and bioinformatic techniques used for the investigation of epigenetic phenomena in plants. Such techniques now allow genome-wide analyses of epigenetic regulation and help to advance our understanding of how epigenetic regulatory mechanisms affect cellular and genome function. To set the scene, we begin with a short background of how the field of epigenetics has evolved, with a particular focus on plant epigenetics. We consider what has historically been understood by the term "epigenetics" before turning to the advances in biochemistry, molecular biology, and genetics which have led to current-day definitions of the term. Following this, we pay attention to key discoveries in the field of epigenetics that have emerged from the study of unusual and enigmatic phenomena in plants. Many of these phenomena have involved cases of non-Mendelian inheritance and have often been dismissed as mere curiosities prior to the elucidation of their molecular mechanisms. In the penultimate section, consideration is given to how advances in molecular techniques are opening the doors to a more comprehensive understanding of epigenetic phenomena in plants. We conclude by assessing some opportunities, challenges, and techniques for epigenetic research in both model and non-model plants, in particular for advancing understanding of the regulation of genome function by epigenetic mechanisms.

**Key words** Epigenetic, Epigenomic, Parent-of-origin, Chromatin, Genetics, Plant science

## 1 Introduction—The Historical Definition of Epigenetics

What does the term epigenetics mean? Over the past decade, a number of excellent reviews have traced the genesis of the term "epigenetics" and its coinage by the Edinburgh-based polymath C. H. Waddington. It is generally agreed that the term arose from his attempts to formulate a model of developmental biology that avoided the reductionism which he considered inherent in the work of the quantitative geneticists which led to the Modern Synthesis (discussed in, e.g., [1, 2]). In 1939, Waddington defined the epigenotype as "the set of organizers and organizing relations to which a certain piece of tissue will be subject during development" [3].

Charles Spillane and Peter C. McKeown (eds.), *Plant Epigenetics and Epigenomics: Methods and Protocols*,
Methods in Molecular Biology, vol. 1112, DOI 10.1007/978-1-62703-773-0_1, © Springer Science+Business Media New York 2014

Specifically, Waddington proposed that the development of a tissue or organism could be conceptualized as occurring along the contours of an "epigenetic landscape" (in: "The Strategy of the Genes" [4]). Following Waddington's work, the field of epigenetics was therefore initially defined in developmental terms and involved the study of the mechanisms by which the genotype brings about the phenotype [1, 5].

The most basic core of Waddington's definition, the need to understand the processes that shape how the information within a genome is regulated in cells and organisms, remains pertinent to modern concepts of epigenetics. However, definitions derived from Waddington's work on developmental landscapes have largely been compromised by the emergence of a more widespread and quite different understanding, originating from advances in molecular biology during the 1980s [1]. More recent definitions of the term epigenetics focus on mechanisms which can change gene expression (or phenotypes) by direct modification of chromosomes while leaving the primary DNA sequence unchanged. In this sense, epigenetics is used in a broad manner to refer to the consequences of DNA and histone modifications (and subsequent chromatin organization). Such definitions may have roots in the less well-known concept of "epigenetics" proposed by Nanney to describe a broad set of "extrachromosomal," "extranuclear heredity," or "functional states" [2, 6]. Although theoretically these gene modification mechanisms can represent an element of the Waddington's developmental model of epigenetics, they lead to a definition of epigenetics with a quite different emphasis. Indeed, it has been argued by Slack that an important consequence of the Waddington's approach to understanding the function of genome during development was a corresponding lack of interest in the molecular details of genetics itself [7].

The use of the same term to refer to two different biological concepts, which are only tangentially related, is hardly ideal. As the developmental biology field can claim historical precedence, while the molecular biology field can claim numerical advantage, the issue seems unlikely to be resolved. Both definitions are however united at certain conceptual levels as both relate to aspects of "soft inheritance" in the sense used by Ernst Mayr so although this volume is concerned with investigations of molecular mechanism, the wider relevance of these for addressing current issues in developmental and evolutionary biology should not be neglected. In fact, examples of "soft inheritance" have played key roles in the discoveries of many molecular epigenetic mechanisms and continue to be the subject of intensive research today. Some consider the possibility that the two interpretations of epigenetics could be reconciled [8]. However, Richards believes this aim to be far-off due to the difficulty of demonstrating heritable components to phenotypic plasticity. The models of "reaction norms" proposed by Woltereck over a century ago [9] and now popularized by Pigliucci and others (*see*, e.g., [10], and below) may provide some opportunity for this. Richards [8] has also drawn attention to studies in species as various as violet, mangrove, wild barley, and diploid potato which have indicated associations between DNA methylation and natural variation for aspects of plant phenotype, although proving a causative relationship remains difficult. If such causative relationships could be established, a model for understanding epigenetic phenomena in the light of Waddingtonian developmental biology could yet prove possible, although the idea that this would represent a move to "post-Darwinian" biology [2] is probably overstated.

## 2 Epigenetics and Molecular Mechanisms

A current day definition of epigenetics is that proposed by Arthur Riggs and colleagues. In this definition, epigenetics is considered to be "the study of mitotically and/or meiotically heritable changes in gene function that cannot be explained by changes in DNA sequence" [11, 12]. Bird [11] further proposed a unifying definition of epigenetic events based on "the structural adaptation of chromosomal regions so as to register, signal, or perpetuate altered activity states." Epigenetic effects can play key roles in "arranging chromosome structure, silencing of tandem repeats and viruses, and expression patterns of genes during development and environmental response" [13] highlighting that genetics and epigenetics are inherently intertwined.

The need to demonstrate that epigenetic heritable changes lead to phenotypic or biologically functional consequences for the organism has led to proposals indicating that "an epigenetic trait is a stably inherited phenotype resulting from changes in a chromosome without alterations in the DNA sequence." [14]. The fact that the heritability of an epigenetic change must be demonstrated is a key feature of any strict definition of epigenetics [15] and is discussed with reference to dosage effects, histone modifications, and transposons in this volume (*see* Chaps. 1, 11, 13, and 14).

The basic components of a DNA-based molecular epigenetic system can be defined by: firstly, (1) a *signal from the environment* that leads to (2) *a responding signal in the cell* that elicits a nongenetic modification of the DNA and (3) a *sustaining signal* that perpetuates the modification through successive cell divisions. At the risk of further adding to the plethora of definitions of epigenetics, it could be considered that the interactions which occur between genomes—during reproduction, hybridization, symbiosis, or interaction with pathogens or viruses—can also be considered as "environmental signals" rather than limiting epigenetic stimuli solely to physical or chemical stimulants such as morphogens or agents of environmental stress.

When seeking to reach an understanding of epigenetics that may serve as a working definition, one important observation made by Bird is that "processes less irrevocable than mutation fall under the umbrella term 'epigenetic' mechanisms" [16] which suggests that epigenetics need not be bound to any particular mechanism. Rather, any change to cellular or organismal function which is heritable but which does not involve permanent changes to the DNA sequence can be classed as epigenetic, again allying this definition to the "soft inheritance" concepts of Mayr. In other organisms, this definition also embraces the inheritance of cellular organization as occurs in Paramecium (and see discussion of maternal effects below) and of prions, which have yet to be demonstrated to exist in plants.

Just as our understanding of what we define as a gene is changing over time [17], the molecular definition of epigenetics is also continuing to develop [1] and suggests that the most comprehensive definitions of epigenetics will continue to extend beyond chromatin mechanisms into considerations of phenotype. Definitions of epigenetics based upon DNA and chromatin modification are the primary focus of this volume, and in particular on highlighting novel methods and approaches that can be used for the epigenetic analysis of DNA and chromatin modifications. Epigenetics research in the "omics" age has the potential to lead to a "systems biology" of epigenetic phenomena where non-Mendelian phenotypic phenomena can be better explained through causal molecular epigenetic mechanisms.

In plants, meiotically heritable phenotypes with an underlying epigenetic basis can be caused by differences in DNA methylation [18], and include changes involving epigenetically modified plant genes such as *peloric*, *rl/bl*, *FWA*, and *SUPERMAN*. In addition, plant research has clearly demonstrated epialleles causing phenotypic changes in epigenetically different lines of *met1* and *ddm1* mutants [19–22] as well as poorly understood effects on development such as the switch to hermaphroditism in *Silene* [23].

Post-translational modifications of histones (e.g., methylation, acetylation, phosphorylation, ubiquitylation, sumoylation) regulate chromatin condensation and accessibility. While histone modifications can clearly elicit different epigenetic states at a locus, in contrast to CG methylation there is scant evidence for meiotically heritable histone modifications [24]. There is also little evidence to support the heritability of different states regarding the three-dimensional organization of genomes in nuclei [25]. In the following section of this chapter, we summarize some examples of epigenetic phenomena for which links between epigenetic mechanisms and phenotype have been established.

## 3 Epigenetic Phenomena and Pioneers of Plant Research

Plant researchers have played a pioneering role in the field of molecular epigenetics, with many of the seminal advances in epigenetics generated using plant models. In many instances, the discovery of a phenomenon in plants has led to the identification of an important epigenetic mechanism. Many of these fall into the class that Goldberg and colleagues spoke of as "numerous biological phenomena, some considered bizarre and inexplicable," and which were "lumped" together as epigenetic [26]. Such phenomena are often characterized by non-Mendelian inheritance, and in several cases have led to the identification of epigenetic mechanisms of general interest for eukaryote biology. Non-Mendelian effects are also associated with inheritance of organellar genomes, including cytoplasmic male sterility [27], although these are not classed as epigenetic under current definitions and are not considered further in this chapter.

The main molecular mechanisms of epigenetic inheritance involve changes to chromatin organization by DNA methylation, histone modification, and the action of chromatin-binding complexes. DNA methylation changes can induce heritable epigenetic changes affecting plant gene regulation and phenotypes. It has been argued that in mammals the H3K27me3-binding Polycomb Group (PcG) complexes are the major chromatin components which are capable of mediating epigenetic transfer of information [28] as it is only these which can generate heritable changes to phenotype. Claims for truly epigenetic (i.e., heritable) effects of chromatin modifiers must therefore be evaluated carefully, and the methods for epigenetic analyses presented in this volume will assist plant researchers to do so.

The discoveries of DNA cytosine methylation, histone modifications, and the panoply of roles for chromatin complexes were predominantly made in yeast and animal systems, e.g., [29–31]. In the case of DNA methylation, initial fundamental insights were largely made in prokaryotes [32] and the universality of these findings, even between prokaryotes and eukaryotes, continues to be increasingly appreciated [33]. On the other hand, plant chromatin has a number of unique features which highlight the molecular diversity of epigenetic mechanisms, and adaptations of likely importance to the sessile, multicellular eukaryotic lifestyle of plants. Gruenbaum and colleagues [34] demonstrated that, in the DNA of Angiosperms, much cytosine methylation occurs in a CHG rather than CG context. This diversity of DNA methylation is linked to the role of small RNAs transcribed by dedicated plant lineage-specific polymerase complexes (Pol IV, Pol V) in triggering heterochromatin formation and represents a major difference between plants and other eukaryotic organisms [35]. Similarly, the diversity of plant histone isoforms has been elegantly shown in a number of studies by Waterborg, e.g. [36], and plant-specific histone modifications and linker histones have also been identified (see http://www.chromdb.org).

An additional distinction of plants derives from their apparent tolerance of mutation. For example, it is possible to knock down a major Arabidopsis methyltransferase such as *MET1* to observe the consequences of loss of DNA methylation throughout a plant life cycle [37], while in mice the equivalent mutation causes mid-gestation embryo lethality [38]. Many studies of animal epigenetic modifiers have been of necessity restricted to cultured cells. The greater tolerance of plants to epigenetic perturbation may have contributed to the many and ongoing epigenetic discoveries made through plant biology research. The following section highlights some of the important advances in epigenetics that have been conducted in plants, including advances involving the discovery of transposable elements and their roles in genome stability; chromatin and its regulation by small RNAs; the non-Mendelian gene regulation observed in paramutation, genomic imprinting, nucleolar dominance, and epigenetic memory systems such as vernalization. In addition there are group of related transgenerational epigenetic effects in plants associated with hybridization, including heterosis and hybrid dysgenesis, in which recent work suggests epigenetic mechanisms may be playing an under-appreciated role.

*Transposable elements.* The discovery of genetic elements capable of generating novel phenotypes by intragenomic mobility by the pioneering scientist Barbara McClintock is a famous narrative for the power of plant genetics. The discovery of transposable elements is also a paradigm for discoveries in epigenetics using plant models, and a good example of a case where a discovery is initially regarded as an oddity, but later found to have much wider relevance. Following the initial discovery of transposons in plants, subsequent mechanistic studies were performed in animal systems, and eventually the development of modern molecular tools allowed a more thorough reassessment of the phenomenon of mobile elements in plants. McClintock discovered that the expression of phenotypes in maize kernels was under the regulation of stochastic genomic components which she termed "controlling elements," and brilliantly surmised that these represented parts of the genome that were capable of transposing from one location in a genome to another location. These elements were later termed transposons or transposable elements (TEs) in recognition of this fact. Transposons are now known to constitute a large and variable proportion of most eukaryotic genomes and to play key roles in genomic and epigenomic evolution [39].

Due to a lack of evidence for any major functional benefits of transposons for cells and organisms, transposable elements have typically been considered as genomic "junk" whereby different transposon classes can be present in many thousands of copies. It was initially suspected [40] and finally proven that there are close relationships between many transposons and viruses, both of which can be considered as classes of genomic parasites. In some instances, transposons have been co-opted into structural or regulatory roles [41, 42]. Both classes of parasite (viruses and transposons) have played key roles in genome evolution [42, 43] and represent key elements involved in epigenetic control of gene regulation and genome integrity [44, 45] (as described further in the discussion of RNAi below).

Epigenetic regulatory mechanisms (e.g., DNA methylation, RNAi, H3K9me2, and H3K4me3) can play a role in ensuring that transposons are kept in a quiescent state, as demonstrated initially by experiments in which disruption of normal DNA methylation patterns can release some transposons from silencing in an apparently stochastic manner [46–50]. Contemporary techniques for analyzing the resumption of activity of transposons and retrotransposons are discussed at greater length in Chap. 13 by DeFraia and Slotkin, and in Chap. 14 by Parisod and colleagues.

*Gene silencing and RNAi.* Perhaps the most significant breakthrough in epigenetics has arisen from the discovery of gene silencing by RNA interference (RNAi), which is now known to be integrally related with the transposable elements and repeat elements originally discovered by McClintock. Plant researchers played a pivotal role in the original discovery of RNAi [51–53], and in subsequent demonstrations that RNAi is essential for correct gene regulation and cellular development throughout the eukaryotes [54].

Downregulation of gene expression by antisense RNA molecules was initially demonstrated in prokaryotes [55], Drosophila [56], Xenopus [57], Dictostylium [58], and plants [59]. The discovery of *Tsix* as an antisense regulator of the mammalian X-inactivation factor *Xist* provided an excellent example of a link between antisense RNA and epigenetic regulation [60]. Although the mechanistic basis initially remained unclear, early investigations demonstrated that antisense RNA had capabilities as a tool for exogenous gene inactivation in animal culture cells [61] and in *C. elegans* [62]. This also revealed the curious fact that suppression of transcripts of endogenous genes by injection of sense, antisense, or dsRNA cognates was heritable [63] and transmissible between cells [64]. Investigation into the mechanisms of this phenomenon, termed RNAi, revealed that it could potentially help explain the enigmatic post-transcriptional gene silencing (PTGS) and co-suppression phenomena which emerged during the initial development of transgenic plants [65], including transgenic plants expressing components of plant viruses in order to elicit pathogen-derived resistance [66].

In the arena of plant virology, homologous interference between RNA viruses was observed as early as the 1970s [67]. Arising from enigmatic findings in plant virology, PTGS was proposed as a plant–pathogen response. From this perspective, Lindbo and Dougherty have reviewed the discovery of RNA-activated sequence-specific RNA degradation, with a particular emphasis on the role played by plant pathology in the discovery of RNAi as a mechanism of PTGS [68]. Significant elements of the discovery of RNAi in plants emerged from initial investigations into pathogen-derived resistance, whereby expression of components of a pathogen's genome (e.g., virus coat protein genes) in a plant cell was employed as a strategy to engineer resistance to the pathogen [69, 70]. While the pathogen-derived resistance designs in this field were typically based on over-expression in transgenic plants cells of a wild-type or mutated protein derived from the viral genome, it became apparent that the resistance elicited to the virus did not require translation of the viral-derived protein [71, 72]. For instance, the Dougherty lab established that the RNA transcript of the coat protein (CP) gene was sufficient to permit resistance to the virus as frameshifted CP genes were effective [70]. The Baulcombe lab demonstrated suppression of virus accumulation in transgenic plants where nuclear genes (with sequence similarity) were subject to gene silencing and proposed a link between the DNA-based transgene methylation and the RNA-based gene silencing process [73].

As in *C. elegans*, dsRNA molecules were found to also be particularly effective for eliciting RNAi in plants [74–76]. Such investigations also led to the proposal that a form of RNA–RNA binding reaction (now known to be mediated by RISC) was responsible for PTGS via transcription of a short RNA signal [77]. These discoveries of homology-dependent gene silencing phenomena in phytopathology and in early transgenic studies were ultimately synthesized with observations from the use of dsRNA as a molecular tool to suggest that RNAi/PTGS was not a response artificially induced by either humans or viruses, but an endogenous mechanism for controlling many nuclear processes by directing chromatin modification in *cis* and in *trans* [78, 79]. While endogenous RNA-silencing pathways are clearly highly conserved and under significant regulatory control, the identification of functional effects of disruption of such pathways remains an active arena of investigation. For instance, a recent report indicates that RNAi is required for transgenerational stability of transposable elements under heat-shock conditions [80, 81].

*Paramutation*. Plant-based research has identified a further example of epigenetic changes having transgenerational effects on gene expression, in the case of paramutation, in which inter-allelic interactions (between paramutagenic and paramutable alleles) can lead to heritable changes in expression of a gene without any change of the underlying DNA sequence. Originally described in pea by William Bateson and Caroline Pellew in 1920 (reviewed [82]), paramutation has since been widely studied in maize (reviewed [83]) and can also be artificially generated as shown in transgenic petunia [84].

Paramutation is an epigenetic phenomenon involving interactions between alleles at a locus, whereby one (paramutagenic) allele can induce a heritable epigenetic change in the expression status of the other (paramutable) homologous allele. Alleles which are not affected by paramutagenic alleles are called neutral alleles. When a paramutated allele is transmitted meiotically, it retains its altered state, even in the absence of a paramutagenic allele in subsequent generations. In some instances, paramutable alleles can become paramutagenic and propagate the effect to other alleles (secondary paramutation), as first described for maize *b1* [85]. The formation of epialleles which exhibit paramutation-like behavior has been shown to occur in response to induced tetraploidy in *Arabidopsis thaliana* polyploids, with implications of induced epigenetic variation for adaptation and evolution of polyploid plants [86].

Although originally considered to be due to somatic pairing, it is now known that paramutation of the maize gene *b1* requires transcription of both DNA strands by an RNA-dependent RNA polymerase [87]. Paramutation is thus related to RNAi in that both phenomena are caused by *trans*-acting RNA molecules produced by plant-specific polymerase complexes and indeed the two RdRPs are closely related [88]. As a further similarity with RNAi, paramutated states are also correlated with DNA methylation [89] and it has been argued that both epigenetic phenomena arose from mechanisms for silencing invasive transposable elements/viruses [90]. As with many other epigenetic phenomena, the first descriptions of paramutation involved discrete "on/off" states (e.g., in relation to gene expression), but more detailed study suggests that many more "partial" cases, involving allelic expression imbalance, also exist [91].

Paramutation is not limited to plants and has also subsequently been demonstrated in animal genomes [92, 93]. For example, paramutation has also been described in mice carrying a mutation in the *Kit* gene [94–98]. The detection of highly penetrant phenotypes from stable paramutable alleles in pigmentation genes of maize may be a harbinger of a pervasive epigenetic surveillance system mediated by RNA [92]. In this context, it is of interest that the machinery which maintains the stability of paramutable alleles is also required for epigenetic control of cell fate-specification leading to sex organ development [99]. Further details of current research into the mechanism of paramutation are provided in Chap. 15 by Vermeersch and colleagues.

*Genomic imprinting.* Genomic imprinting is a phenomenon observed in flowering plants and mammals, in which a locus is differentially expressed depending on whether the allele is inherited maternally or paternally [100, 101]. The term genomic imprinting was first used to describe the elimination of paternal chromosomes during spermatogenesis in Sciarid flies [102]. However, genomic imprinting at the gene level was first demonstrated by Kermicle in 1979 for the maize R locus [103]. The discovery of gene-specific imprinting in maize was followed in 1984 by the discovery of genomic imprinting in mammals in a series of pronuclear transplantation experiments involving androgenetic and gynogenetic diploid embryos [104, 105]. Imprinting is now clearly involved in a wide range of human medical conditions, and large numbers of imprinted loci have now been described in humans and other mammals [106].

Recent advances in next-generation sequencing technology have allowed a more thorough description of the extent of genomic imprinting in plants, laying the basis for critical assessments of a range of theories for how and why imprinting evolved [100, 101, 103]. The parent-of-origin-specific uniparental expression of imprinted plant genes is due to alleles of certain genes being modified by "epigenetic" marks during male and female gametogenesis, whereby the altered epigenetic state at the locus (e.g., expression level) persists after fertilization [100, 107]. Genomic imprinting in plants is considered to predominantly affect genes in the endosperm [107–109]. To date, only a small number of imprinted genes have been shown to be essential or important for endosperm development in plant seed, e.g., *MEDEA* in *Arabidopsis thaliana* [110]. Misregulation of some imprinted genes has also been shown to lead to seed abortion in hybrid and polyploid crosses [111], possibly due to dosage effects of the kind discussed in Chap. 2 by Birchler and Veitia.

The use of RNA-Seq (and cDNA-AFLP) has recently allowed the identification of large numbers of candidate imprinted genes (both maternally expressed imprinted genes iMEGs and paternally expressed iPEGs) in the monocot crops *Zea mays* and *Oryza sativa* [112, 113] and in the model eudicot *Arabidopsis thaliana* [114–116]. At present, there is little consensus on why and how imprinting evolved (in either mammals or plants), with a range of theories ranging from parental conflict between maternal and paternal genes for maternal resource allocation [117], to being a mechanism for control of dosage-sensitive genes. The techniques that can be employed to identify and validate imprinted plant genes are described in further detail in Chap. 5 by Day and Macknight, and from work in our laboratory in Chap. 6.

*Nucleolar dominance.* Nucleolar dominance is a particular form of uniparental gene expression frequently observed in interspecific hybrids (of both plants and animals) in which only the ribosomal RNA (rRNA) genes from one of the parental species will be actively transcribed, while those derived from the other parent will be silenced [118]. The rRNA genes are usually present in tandem repeats at loci which are termed nucleolar organizer regions (NORs) because they give rise to the nucleolus when transcribed, the link between these loci and the nucleolus being another seminal discovery in cell biology made by Barbara McClintock in maize [119]. As only the rRNA genes of the dominant parent participate in the assembly of the nucleolus in hybrids, this phenomenon is termed "nucleolar dominance." Nucleolar dominance differs from genomic imprinting in that the NORs from the dominant parental species are dominant regardless of the cross direction. Nucleolar dominance was originally discovered through research into plant reproduction involving interspecific crosses in the genus Crepis (hawk's-beard) published in 1934 by Navashin. He coined the term "differential amphiplasty" to describe his observation of uniparental changes to the chromosomes with which mitotic strictures known as "secondary constrictions" were associated (reviewed [120]). In the same year, McClintock determined the role of the NOR in generating the nucleolus and showed that Navashin's differential amphiplasty represented an interaction between the NORs of different species which could be organized as a simple hierarchy of dominance [119].

As with paramutation and imprinting, nucleolar dominance as an epigenetic phenomenon was subsequently discovered to occur in animals as well as plants, and it was in animals that much of the subsequent molecular characterization of nucleolar dominance was performed. It was, for example, through research in interspecific hybrids of *Xenopus* that the term "nucleolar dominance" first emerged, and in the *Xenopus* system that the links between the cellular effects and the activation or repression of tandem repeats of rRNA genes was elucidated [121].

In recent years, nucleolar dominance has elicited more widespread interest as a model for the differences between active and inactive eukaryotic genes [122, 123], with the advantage that the differences between arrays of rDNA genes can be visualized by microscopy at the karyotypic level. The use of amphidiploid hybrids between different species of the Arabidopsis genus has played a key role in recent research into the epigenetic mechanisms of nucleolar dominance. Models for nucleolar dominance initially considered genetic mechanisms based on competition for transcription factors, along the lines which McClintock herself had proposed [119]. However, the involvement of DNA methylation, proposed on the basis of work in wheat [124], and of histone modifications in Arabidopsis [125], established that nucleolar dominance is primarily an epigenetic effect. Psoralen cross-linking experiments on chromatin indicated that the difference between active and inactive rRNA genes in a pure-bred organism was also chromatin-dependent and may be a manifestation of similar mechanisms regulating nucleolar dominance in hybrids [126]. Recent work in non-hybrid

*Arabidopsis thaliana* suggests that control of rRNA gene expression is affected by natural variation [127], chromatin-modifying enzymes, [128] and histone and DNA modifications [129, 130] in a similar manner as nucleolar dominance in hybrids. Finally, recent work has indicated that these regulatory systems may make use of small RNA intermediaries which permit the cell to distinguish between different rRNA populations [131].

*Flowering time and the epigenetic memory of winter.* Vernalization is the process whereby exposure to a period of cold (e.g., as occurs over the course of a winter) is required prior to a plant undergoing the transition from the vegetative to the reproductive growth phase [132, 133]. Vernalization requirements of different plant species and varieties (e.g., spring vs. winter cereals) would have been known to farmers and the earliest plant breeders. Whyte and Hudson [134] quote evidence suggesting that the vernalization requirements of winter wheat were already being exploited in the 1830s. Gustav Gassner was one of the first who attempted systematic study of vernalization in 1918. Amasino argues that the principle importance of Gassner's work was to demonstrate the *generality* of vernalization across many plant families [135]. The advocacy by Trofim Lysenko in the 1930s of a politicized version of vernalization (c.f. Jarovization, derived from the Russian for "spring crops"; [136]) as a mechanism for introducing near-immediate acquisition of heritable cold-hardiness, provides an extreme example of the risks of muddying political theories with naive scientific beliefs and exerted a major negative impact on Soviet genetics for decades [137]. This culminated in the Politburo forbidding all research into Mendelian genetics in 1948, apparently with the personal imprimatur of Joseph Stalin. This sad episode in the history of plant biology hinges around the distinction between an epigenetic phenomenon being mitotically heritable through the life-span of an organism but not meiotically transmissible to the offspring. At the cellular level, the "memory" of cold-induced vernalization is transmissible through mitosis within the growing meristem, but is not transmissible through meiosis and subsequent gametes to the next generation.

The genetic basis of vernalization has now been partially unravelled in *Arabidopsis thaliana* and has revealed that environmental and developmental cues are integrated via four pathways which converge upon the genes *FRI* and *FLC* [132, 138]. FLC is epigentically silenced by vernalization (cold treatment) and acts by reducing the protein levels of three promoters of flowering, FT, FD, and SOC1. Both *FLC* and *FT* are regulated by epigenetic factors involving DNA methylation and a range of histone modifications [139, 140]. A role for epigenetic mechanisms in the control of flowering in *Arabidopsis thaliana* was first suggested by the identification of the vernalization inducible *VIN3*, encoding a PHD finger protein which acts as an upstream regulator of FLC (in association with PRC2 complex) by affecting histone modification at the *FLC* locus [141].

As the requirement for vernalization has evolved multiple times in multiple plant taxa, this suggests that preexisting epigenetic pathways may be repeatedly adapted for roles in transmitting cellular memories of the environment. Amasino has suggested that acquisition of vernalization requirements could have occurred independently in many plant lineages if they evolved in the tropics before radiating into climatic zones with more severe winters [135]. Finally, recent work has identified a role for a long noncoding RNA (lncRNA) in recruiting the PcG complexes which contain VIN3 and silence *FLC* [142], highlighting a further role for RNA which parallels its functions in other systems of epigenetic regulation. As RNA molecules have many characters desirable in temperature sensors, similar lncRNAs may also affect vernalization and other temperature responses in other plant species. A case study of the analysis of chromatin modifications in controlling key components of vernalization in plants is described in Chap. 11 of this volume, by Song and colleagues.

*Inbreeding, heterosis, and hybrid dysgenesis.* Inbreeding frequently evolves in plant lineages, perhaps "more often than any other adaptation" [143]. A curious fact is that many plants maintain a "mixed mating" system despite this being predicted to be an evolutionarily unstable situation [144]. The inbreeding "selfing syndrome" is often found to cause a loss of viability, fertility, and overall fitness, termed as inbreeding depression. Inbreeding depression may be partly explained by epigenetic changes affecting the organism, as endogenous removal of aberrant DNA methylation which accumulates following selfing of *Scabiosa columbaria* reverses the associated inbreeding depression [145]. Biemont has argued in favor of an epigenetic component to inbreeding depression, particularly on the basis of studies of Arabidopsis epiRIL populations [146]. These were recombinant inbred lines formed by crossing wild-type plants to DNA methylation mutants *met1-3* and *ddm1*, in the same genetic background [48].

Inbreeding depression is of particular interest due to its involvement in the phenomenon of heterosis, in which inbred lines of (usually outcrossing) organisms are crossed to generate F1 hybrid progeny with a higher fitness score than the average of their parents (midpoint heterosis) or the highest score parent (best-parent heterosis). Heterosis involves changes to both gene expression and nuclear organization, suggesting that hybridization between different strains and species of organisms has phenotypic consequences for an organism well beyond the possibility of nucleolar dominance (see above). Heterosis has been reported in agricultural practices since antiquity, and was of major importance for twentieth century agriculture following its use to increase yield and other beneficial phenotypes particularly in the US maize crop [147]. Because heterosis is generated from the crossing of two inbred lines, heterosis may be considered as the result of relieving inbreeding depression. Indeed, hybrid dysgenesis which occurs from crossing of "over-diverged" lines may also be considered as the "opposite" form of heterosis (see below). Recent studies of ovule number in reciprocal F1 hybrid triploid plants in Arabidopsis thaliana have revealed both hybrid dysgenesis and hybrid advantage (heterosis) effects [148], while it has also been shown that genomic dosage is a contributor to heterosis effects in F1 hybrid triploids of maize [149]. Although the term heterosis is usually considered to refer to the transgressive phenotypes of artificially bred F1 hybrid plants, related phenotypes are also observed in naturally occurring hybrids (and see also the review of Birchler and Veitia in Chap. 2 of this volume), especially of plants which tolerate hybridity well even to the extent of forming natural triple hybrids in a few exceptional cases [150].

Heterosis/hybrid vigor was a major focus of both plant breeders and theoretical geneticists throughout the twentieth century (a historical view is provided by [151]). A key question under investigation is whether heterosis is predominantly resulting from dominance or overdominance. The possible roles of epigenetic mechanisms in heterosis have not received as much attention, even though dominance effects may have an underlying gene expression level basis that could be controlled by DNA methylation and histone modifications. Indeed, DNA methylation and histone modifications are found to alter in F1 hybrid rice, and to correlate with altered transcription of parental alleles [152]. In addition, it has been concluded that much of the transgressive inheritance of expression levels in rice F1 hybrids is likely caused by epimutations and trans-effects [153]. Furthermore, changes in the levels of 24 nt siRNAs in F1 hybrids compared to parental lines have been observed in *Arabidopsis thaliana* [154].

Clearly, the crossing of divergent germplasm does not always generate heterosis and may instead induce so-called hybrid dysgenesis. This is typically the case if the parental lines are too distantly related. If fertilization is able to occur under these circumstances, F1 seed abortion typically results. Hybrid dysgenesis may be considered an opposite effect from that of heterosis and may also have an epigenetic component. In the most extreme case, the result may be a full reproductive barrier with the potential to act as a speciation mechanism (see below). In the context of hybrids, consideration should also be given to the importance of allopolyploidy [155–160]. At a population level, allopolyploidy may stabilize hybrid gene expression [161], and make hybrids ecologically viable by allowing them undergo correct pairing at meiosis (although hybrids can also become successful through reproducing vegetatively or via apomixis). Allopolyploidy is now known to induce major effects on DNA methylation, as well as on small RNA expression, nucleosome positioning, histone modification and overall nuclear organization. Recent data suggests similar effects might also perturb gene expression in autopolyploids which lack hybridity, although no mechanism has been identified for this [162].

Heterosis and the associated phenomena of inbreeding depression and hybrid dysgenesis have been extensively studied in a multitude of plant systems, despite which their molecular explanations often remain elusive [163, 164]. While all three phenomena have genetic components, each may also be affected by the epigenetic consequences of both hybridization and gene dosage effects on F1 hybrid genome functioning [165]. The "omics" era has heralded new opportunities for unravelling the mechanistic basis of heterosis and allied phenomena where genomic, epigenomic, transcriptomic, proteomic, and metabolomic technical advances provide the necessary tools. For example, the expression changes observed in F1 hybrids can be amenable to further analysis by RNA-Seq techniques, as used by Ng et al. (and reviewed by Ng and colleagues in Chap. 3 of this volume).

## 4 Innovations for Improved Understanding of Plant Epigenetics and Epigenome Dynamics

Clearly, plant biology research has played major roles in advancing our understanding of epigenetics, and provides the background to the development of the improved techniques necessary for generating ongoing advances in plant epigenetics and epigenomics. This book provides a range of chapters which describe robust protocols for plant epigenetics and epigenomics research. While many of the chapters in the book reflect the importance of *Arabidopsis thaliana* and its relatives as a tool for plant epigenetic research, the methods provided are typically applicable to any of the growing numbers of plant species with sequenced genomes (see http://www.phytozome.net).

The volume begins with chapters by Birchler and Veitia (*see* Chap. 2) and Ng et al. (*see* Chap. 3) on polyploidy (genome dosage, gene balance hypothesis, allopolyploidy) and hybridization, phenomena which alter the epigenetic context of plant genomes on a scale seldom observed in other taxa. The advent of next-generation sequencing technologies has accelerated the rate of discovery in molecular biology and is having profound impacts on our ability to conduct epigenetic and epigenomic analyses in plants [166, 167]. In Chap. 3, Ng et al. demonstrate how RNA-seq can be used for transcriptome analysis that can resolve allele-specific or homeologue-specific expression patterns in allotetraploid hybrids. To accurately detect cases of allele-specific expression within RNA-seq datasets, robust bioinformatics procedures are necessary in order to avoid any inadvertent biases or inaccurate interpretations. Hence, in Chap. 4 Korir and Seoighe provide some caveats to consider regarding such analysis and provide a pipeline that can be applied for robust detection of cases of allele-specific expression in RNA-seq datasets. Such techniques are of particular significance for genome-wide transcriptome analysis of genomic imprinting and other dosage effects in hybrids. While RNA-seq is extremely powerful for detection of allele-specific expression on a genome-wide scale, complementary techniques which can be employed at the individual locus level are also necessary (for both validation and more focussed investigations). Hence, Chaps. 5 (by Day and Macknight) and 6 (by McKeown et al.) in this volume provide details of two techniques (high resolution melt-curve analysis and pyrosequencing) which can be used at the individual locus level in plants to determine the extent of allele-specific expression at a locus.

Unlike the DNA genome which is largely hardwired, the epigenome can change dramatically in a spatiotemporal manner during the development of any multicellular organism. As a result, the epigenome of different developmental stages, organs, tissues, and cells can be dramatically different. Hence, there is a potential danger of misinterpretation of epigenomic changes when samples from multiple cells or tissue types contain multiple epigenome signatures. To unravel this level of complexity, it is necessary to have techniques that allow for isolation of particular cell-types in plants so that they can be subject to transcriptome and epigenome analyses. Chapter 7 by Weinhofer and Köhler provide a method that allows for fluorescence-activated cell sorting (FACS) to be deployed for cell type-specific RNA and chromatin profiling of plant endosperm tissue. In addition to destructive analysis of plant samples, live-imaging techniques are also necessary which can focus on specific biological processes (e.g., gametogenesis and double fertilization) including how such processes are perturbed in epigenetic modifier backgrounds (*see* Chap. 8 by Ingouff).

The Encyclopedia of DNA Elements (ENCODE) project on the human genome has demonstrated the power of systematically mapping regions of transcription, transcription factor association, chromatin structure, and histone modification across the genome [168]. The vast majority of techniques employed within the ENCODE project can equally be applied in plants so that a more comprehensive (and integrated) understanding of the genome and epigenome dynamics of relevance to plant growth and development can be obtained. In Chaps. 8–11 in this volume, a range of plant epigenome methods are presented which allow for genome-wide or locus-directed analysis of plant chromatin modifications concerning DNA methylation (*see* Chaps. 8 and 9) or histone modifications (*see* Chaps. 10 and 11). Chapter 9 by Cortijo et al. provides a powerful method for genome-wide analysis of DNA methylation in plants through use of immunoprecipitation of methylated DNA followed by hybridization to genome tiling arrays (MeDIP-chip). In addition, the use of MS-AFLP for assaying DNA methylation in a non-model species is described by Albertini and Marconi in Chap. 10. The panoply of histone modifications that can elicit differential chromatin states requires robust techniques for detection of different histone modifications in plant genomes. In Chap. 11, Song et al. describe a robust chromatin immunoprecipitation (ChIP) method for *Arabidopsis thaliana* which can be adopted for other plant species, and which is compatible with multiple downstream applications including qPCR, tilling arrays, and high-throughput sequencing. Luo and Lam provide in Chap. 12 a methodology for quantitative ChIP-seq where ChIP is interfaced with next-generation RNA sequencing using the SOLiD™ 2.0 high-throughput sequencing platform.

Chapters 12 and 13 are concerned with the interplay between transposons and epigenetic regulation [49]. Chapter 13 by DeFraia and Slotkin provides a detailed account of the range of techniques that can be employed to analyze the progress and epigenetic regulation of LTR retrotransposons through their replication cycle in plants. In addition, in Chap. 14 Parisod et al. provide methodologies for the use of sequence-specific amplified polymorphism (SSAP) and the related methyl-sensitive transposon display (MSTD) for analysis of the genome and epigenome dynamics of transposable elements in plant genomes. The final Chap. 15 by Vermeersch et al. highlights the epigenetic phenomena of transitive gene silencing in plants and how transitive silencing assays can be employed for investigation of this epigenetic phenomenon.

Epigenomics can risk offering a "one-dimensional" view of the cell [169] in cases where, for example, a single type of chromatin modification is viewed in isolation. To combat this, the techniques described in this volume engender possibilities for generating more integrated data that can be combined with other datasets to provide the basis for a more systems epigenomics approach to biological questions. Such datasets would likely include genomic, proteomic, and metabolomic studies, analyses of small RNA, in addition to the tools of cell and developmental biology. To integrate large epigenomic datasets will however require the development of community standards to allow valid comparisons between experiments. These could perhaps be modelled upon the MIAME guidelines for interpretation of microarray data. Possible guidelines for ChIP-seq have been developed for use in different human culture cells by, for example, the ENCODE consortium (https://www.genome.gov/10005107), the NIH Roadmap Epigenomics Mapping Consortium (REMC, http://www.roadmapepigenomics.org/), and the Beta Cell Biology Consortium (http://www.betacell.org/), amongst others. As yet, these remain at an early stage and no plant-specific guidelines have been agreed upon.

The specific biological questions which the types of techniques described in this volume may be used to investigate could be divided into "bottom-up" questions (determining the functions of the chromatin patterns revealed by next-generation sequencing approaches) and "top-down" questions (determining the mechanisms of epigenetic phenomena which remain unexplained). So far, genome-wide chromatin analysis techniques have largely been restricted to large cell populations following different differentiation pathways, and representing an average of what may be very different individual patterns. For a counter-example, see the analysis of endosperm chromatin by cell-sorting in Chap. 7, by Weinhofer and Köhler. In this context, the application of RNA-seq to individual cells holds great promise for more accurately dissecting the roles of chromatin changes during cell differentiation. From the perspective of understanding the heritability of chromatin organization, alluring targets will be the stem cells in the plant meristem [170] and the developing gametes. Comparisons between these may allow valuable conclusions to be drawn about the possibilities of epigenetic inheritance during mitosis and meiosis which is essential for advancing our understanding of the functional, developmental, ecological, and evolutionary significances of epigenetic changes in multicellular organisms.

The plant epigenomic analysis approaches in this volume can help us to understand the mechanics and diversity of differential chromatin effects across the genome and how these relate to phenotypes displaying epigenetic effects. To interpret the biological significance of large datasets generated by "omics" or genome-wide approaches poses many challenges in terms of data-handling (see the descriptions in Chaps. 3 and 8, on ASE and MeDIP, in particular), and require robust methodologies to distinguish which data-points are biologically significant from those which constitute noise. In other words, there is a need for robust experimental means of determining whether epigenetic marks are causal for changes in gene or cellular function, rather than being downstream consequences. To achieve this will require a renewed appreciation of the phenotypes associated with changes at the chromatin level which could include aspects of cell fate, differentiation patterns, plant physiology, plant fitness, and environmental responses. Ironically, understanding the mechanisms by which genes and gene products interact under genome and epigenome regulation may go some way towards defining the "epigenetic landscape" of cellular response during development that Waddington originally envisaged. The techniques in this volume will also be instrumental in advancing our understanding of the biological significance of different epigenetic phenomena, including to what extent heritability of different epigenetic states can play any roles at the organismal, evolutionary, and ecosystem levels.

## 5 Conclusion

Epigenetics is one of the most exciting areas of investigation in biology and has remained so over many decades. This opening chapter has highlighted a continuum of contributions from the plant epigenetics community to the current-day understanding of epigenetics and epigenome regulation. For some key epigenetic phenomenon, we have described the prominent role of plant research in its study. Such contributions span decades of investigations by past and extant pioneers of plant epigenetics research. Discoveries in plants have proved instrumental in revealing processes now known to be essential to our understanding of fundamental eukaryote biology. As our understanding of epigenetics continues to develop, it can be expected that plants will continue to prove their worth as basic and applied models for epigenetics research. Indeed, given the reliance of human society on plants for our existence, it is imperative that we deepen our understanding of the role of epigenetics in plant form, function, adaptation, and evolution.

### References

1. Haig D (2004) The (dual) origin of epigenetics. Cold Spring Harb Symp Quant Biol 69:67–70
2. Huang S (2012) The molecular and mathematical basis of Waddington's epigenetic landscape: a framework for post-Darwinian biology? Bioessays 34:149–157
3. Waddington CH (1939) An introduction to modern genetics. Allen and Unwin, London
4. Waddington CH (1957) The strategy of the genes. Allen and Unwin, London
5. Waddington CH (1942) Canalization of development and the inheritance of acquired characters. Nature 150:563–565
6. Nanney DL (1958) Epigenetic control systems. Proc Natl Acad Sci U S A 44:712–717
7. Slack JM (2002) Conrad Hal Waddington: the last renaissance biologist? Nat Rev Genet 3:889–895
8. Richards EJ (2011) Natural epigenetic variation in plant species: a view from the field. Curr Opin Plant Biol 14:204–209
9. Woltereck R (1909) Weitere experimentelle undersuchungen über Artveranderung, speziell über das Wesen quantitativer Artunterscheide bei Daphnien. Verhandlungen der Deutschen Zoologischen Gesellschaft 19: 110–173
10. Pigliucci M (2007) Do we need an extended evolutionary synthesis? Evolution 61:2743–2749
11. Bird A (2007) Perceptions of epigenetics. Nature 447:396–398
12. Russo VEA, Martienssen RA, Riggs AD (eds) (1996) Epigenetic mechanisms of gene regulation. Cold Spring Harbor Laboratory Press, Woodbury
13. Grant-Downton RT, Dickinson HG (2006) Epigenetics and its implications for plant biology 2. The 'epigenetic epiphany': epigenetics, evolution and beyond. Ann Bot 97:11–27
14. Berger SL, Kouzarides T, Shiekhattar R, Shilatifard A (2009) An operational definition of epigenetics. Genes Dev 23:781–783
15. Jaenisch R, Bird A (2003) Epigenetic regulation of gene expression: how the genome integrates intrinsic and environmental signals. Nat Genet 33:S245–S254
16. Bird A (2002) DNA methylation patterns and epigenetic memory. Genes Dev 16:6–21
17. Gerstein MB, Bruce C, Rozowsky JS, Zheng D, Du J, Korbel JO, Emanuelsson O, Zhang ZD, Weissman S, Snyder M (2007) What is a gene, post-ENCODE? History and updated definition. Genome Res 17:669–681
18. Hauser M-T, Aufsatz W, Jonak C, Luschnig C (2011) Transgenerational epigenetic inheritance in plants. Biochim Biophys Acta 1809: 459–468
19. Kakutani T, Jeddeloh JA, Flowers SK, Munakata K, Richards EJ (1996) Developmental abnormalities and epimutations associated with DNA

hypomethylation mutations. Proc Natl Acad Sci U S A 93:12406–12411

20. Soppe WJJ, Jacobsen SE, Alonso-Blanco C, Jackson JP, Kakutani T, Koornneef M, Peeters AJM (2000) The late flowering phenotype of *fwa* mutants is caused by gain-of-function epigenetic alleles of a homeodomain gene. Mol Cell 6:791–802
21. Rangwala SH, Elumalai R, Vanier C, Ozkan H, Galbraith DW, Richards EJ (2006) Meiotically stable natural epialleles of Sadhu, a novel Arabidopsis retroposon. PLoS Genet 2:e36
22. Johannes F, Porcher E, Teixeira FK, Saliba-Colombani V, Simon M, Agier N, Bulski A, Albuisson J, Heredia F, Audigier P, Bouchez D, Dillmann C, Guerche P, Hospital F, Colot V (2009) Assessing the impact of transgenerational epigenetic variation on complex traits. PLoS Genet 5:e1000530
23. Janoušek B, Široký J, Vyskot B (1996) Epigenetic control of sexual phenotype in a dioecious plant, *Melandrium album*. Mol Gen Genet 250:483–490
24. Pecinka A, Mittelsten Scheid O (2012) Stress-induced chromatin changes: a critical view on their heritability. Plant Cell Physiol 53: 801–808
25. Rapp RA, Wendel JF (2005) Epigenetics and plant evolution. New Phytol 168:81–91
26. Goldberg AD, Allis CD, Bernstein E (2007) Epigenetics: a landscape takes shape. Cell 128:635–638
27. Chase CD (2007) Cytoplasmic male sterility: a window to the world of plant mitochondrial–nuclear interactions. Trends Genet 23:81–90
28. Deaton AM, Bird A (2011) CpG islands and the regulation of transcription. Genes Dev 25:1010–1022
29. Bird AP (1986) CpG-rich islands and the function of DNA methylation. Nature 321:209–213
30. Doerfler W (1983) DNA methylation and gene activity. Annu Rev Biochem 52:93–124
31. Wigler MH (1981) The inheritance of methylation patterns in vertebrates. Cell 24: 285–286
32. Bestor TH, Verdine GL (1994) DNA methyltransferases. Curr Opin Cell Biol 6:380–389
33. Youngson NA, Chong S, Whitelaw E (2011) Gene silencing is an ancient means of producing multiple phenotypes from the same genotype. Bioessays 33:95–99
34. Gruenbaum Y, Navehmany T, Cedar H, Razin A (1981) Sequence specificity of methylation in higher plant DNA. Nature 292:860–862
35. Lahmy S, Bies-Etheve N, Lagrange T (2010) Plant-specific multisubunit RNA polymerase in gene silencing. Epigenetics 5:4–8
36. Waterborg JH (1990) Sequence analysis of acetylation and methylation in two histone H3 variants of alfalfa. J Biol Chem 265: 17157–17161
37. Finnegan EJ, Peacock WJ, Dennis ES (1996) Reduced DNA methylation in *Arabidopsis thaliana* results in abnormal plant development. Proc Natl Acad Sci U S A 93: 8449–8454
38. Li E, Bestor TH, Jaenisch R (1992) Targeted mutation of the DNA methyltransferase gene results in embryonic lethality. Cell 69: 915–926
39. Lisch D (2012) How important are transposons for plant evolution? Nat Rev Genet 14:49–61
40. Shimotohno K, Mizutani S, Temin HM (1980) Sequence of retrovirus provirus resembles that of bacterial transposable elements. Nature 285:550–554
41. Coen ES, Carpenter R, Martin C (1986) Transposable elements generate novel spatial patterns of gene expression in *Antirrhinum majus*. Cell 47:285–296
42. Bennetzen JL (2005) Transposable elements, gene creation and genome rearrangement in flowering plants. Curr Opin Genet Dev 15: 621–627
43. McDonald JF (1995) Transposable elements: possible catalysts of organismic evolution. Trends Ecol Evol 10:123–126
44. Lippman Z, Gendrel A-V, Black M, Vaughn MW, Dedhia N, Richard McCombie W, Lavine K, Mittal V, May B, Kasschau KD, Carrington JC, Doerge RW, Colot V, Martienssen R (2004) Role of transposable elements in heterochromatin and epigenetic control. Nature 430:471–476
45. Parisod C, Salmon A, Zerjal T, Tenaillon M, Grandbastien M-A, Ainouche M (2009) Rapid structural and epigenetic reorganization near transposable elements in hybrid and allopolyploid genomes in Spartina. New Phytol 184:1003–1015
46. Hirochika H, Okamoto H, Kakutani T (2000) Silencing of retrotransposons in Arabidopsis and reactivation by the *ddm1* mutation. Plant Cell 12:357–368
47. Okamoto H, Hirochika H (2001) Silencing of transposable elements in plants. Trends Plant Sci 6:527–534
48. Reinders J, Wulff BBH, Mirouze M, Marí-Ordóñez A, Dapp M, Rozhon W, Bucher E, Theiler G, Paszkowski J (2009) Compromised stability of DNA methylation and transposon immobilization in mosaic Arabidopsis epigenomes. Genes Dev 23:939–950
49. Lisch D (2009) Epigenetic regulation of transposable elements in plants. Annu Rev Plant Biol 60:43–66

50. Cui X, Jin P, Gu L, Lu Z, Xue Y, Wei L, Qi J, Song X, Luo M (2013) Control of transposon activity by a histone H3K4 demethylase in rice. Proc Natl Acad Sci U S A 110: 1953–1958
51. Bots M, Maughan S, Nieuwland J (2006) RNAi Nobel ignores vital groundwork on plants. Nature 443:906
52. Jorgensen R (2006) Plants, RNAi, and the Nobel prize. Science 314:1242–1243
53. Matzke M, Matzke AJM (2006) Plants, RNAi, and the Nobel prize. Science 314:1242
54. Cibrián-Jaramillo A, Martienssen RA (2009) Darwin's "abominable mystery": the role of RNA interference in the evolution of flowering plants. Cold Spring Harb Symp Quant Biol 74:267–273
55. Green PJ, Pines O, Inouye M (1986) The role of antisense RNA in gene regulation. Annu Rev Biochem 55:569–597
56. Rosenberg UB, Preiss A, Seifert E, Jäckle H, Knipple DC (1985) Production of phenocopies by Krüppel antisense RNA injection into drosophila embryos. Nature 313:703
57. Harland R, Weintraub H (1985) Translation of mRNA injected into Xenopus oocytes is specifically inhibited by antisense RNA. J Cell Biol 101:1094–1099
58. Crowley TE, Nellen W, Gomer RH, Firtel RA (1985) Phenocopy of discoidin I-minus mutants by antisense transformation in Dictyostelium. Cell 43:633
59. Cornehssen M, Vandewiele M (1989) Both RNA level and translation efficiency are reduced by anti-sense RNA in transgenic tobacco. Nucleic Acids Res 17:833–843
60. Lee JT, Davidow LS, Warshawsky D (1999) Tsix, a gene antisense to Xist at the X-inactivation centre. Nat Genet 21: 400–404
61. Izant JG, Weintraub H (1984) Inhibition of thymidine kinase gene-expression by antisense RNA—a molecular approach to genetic analysis. Cell 36:1007–1015
62. Guo S, Kemphues KJ (1995) par-1, a gene required for establishing polarity in C. elegans embryos, encodes a putative Ser/Thr kinase that is asymmetrically distributed. Cell 81: 611–620
63. Grishok A, Mello CC (2002) RNAi (nematodes: *Caenorhabditis elegans*). Adv Genet 46:339–360
64. Fire A, Xu SQ, Montgomery MK, Kostas SA, Driver SE, Mello CC (1998) Potent and specific genetic interference by double-stranded RNA in *Caenorhabditis elegans*. Nature 391:806–811
65. Napoli C, Lemieux C, Jorgensen R (1990) Introduction of a chimeric chalcone synthase gene into petunia results in reversible co-suppression of homologous genes in trans. Plant Cell 2:279–289
66. Abel PP, Nelson RS, De B, Hoffmann N, Rogers SG, Fraley RT, Beachy RN (1986) Delay of disease development in transgenic plants that express the tobacco mosaic virus coat protein gene. Science 232:738
67. Kassanis B, White RF (1972) Interference between two satellite viruses of tobacco necrosis virus. J Gen Virol 17:177–183
68. Lindbo JA, Dougherty WG (2005) Plant pathology and RNAi: a brief history. Annu Rev Phytopathol 43:191–204
69. Sanford JC, Johnston SA (1985) The concept of parasite-derived resistance—deriving resistance genes from the parasites own genome. J Theor Biol 113:395–405
70. Lindbo JA, Silva-Rosales L, Proebsting WM, Dougherty WG (1993) Induction of a highly specific antiviral state in transgenic plants: implications for regulation of gene expression and virus resistance. Plant Cell 5: 1749–1759
71. Smith HA, Swaney SL, Parks TD, Wernsman EA, Dougherty WG (1994) Transgenic plant virus resistance mediated by untranslatable sense RNAs: expression, regulation, and fate of nonessential RNAs. Plant Cell 6: 1441–1453
72. Mueller E, Gilbert J, Davenport G, Brigneti G, Baulcombe DC (2002) Homology-dependent resistance: transgenic virus resistance in plants related to homology dependent gene silencing. Plant J 7:1001–1013
73. English JJ, Mueller E, Baulcombe DC (1996) Suppression of virus accumulation in transgenic plants exhibiting silencing of nuclear genes. Plant Cell 8:179–188
74. Montgomery MK, Fire A (1998) Double-stranded RNA as a mediator in sequence-specific genetic silencing and co-suppression. Trends Genet 14:255–258
75. Metzlaff M, O'dell M, Cluster PD, Flavell RB (1997) RNA-mediated RNA degradation and chalcone synthase A silencing in petunia. Cell 88:845–854
76. Dalmay T, Hamilton A, Rudd S, Angell S, Baulcombe DC (2000) An RNA-dependent RNA polymerase gene in *Arabidopsis* is required for posttranscriptional gene silencing mediated by a transgene but not by a virus. Cell 101:543–553
77. Dougherty WG, Parks TD (1995) Transgenes and gene suppression—telling us something new. Curr Opin Cell Biol 7:399–405
78. Sen GL, Blau HM (2006) A brief history of RNAi: the silence of the genes. FASEB J 20: 1293–1299
79. Baulcombe D (2004) RNA silencing in plants. Nature 431:356–363

80. Ito H, Gaubert H, Bucher E, Mirouze M, Vaillant I, Paszkowski J (2011) An siRNA pathway prevents transgenerational retrotransposition in plants subjected to stress. Nature 472:115–119
81. Staiger D, Korneli C, Lummer M, Navarro L (2012) Emerging role for RNA-based regulation in plant immunity. New Phytol 197(2): 394–404
82. Brink RA (1973) Paramutation. Annu Rev Genet 7:129–152
83. Chandler VL, Stam M (2004) Chromatin conversations: mechanisms and implications of paramutation. Nat Rev Genet 5:532–544
84. Meyer P, Heidmann I, Niedenhof I (1993) Differences in DNA methylation are associated with a paramutation phenomenon in transgenic *Petunia*. Plant J 4:89–100
85. Coe EH (1966) Properties origin and mechanism of conversion-type inheritance at b locus in maize. Genetics 53:1035–1063
86. Mittelsten Scheid O, Afsar K, Paszkowski J (2003) Formation of stable epialleles and their paramutation-like interaction in tetraploid *Arabidopsis thaliana*. Nat Genet 34: 450–454
87. Alleman M, Sidorenko L, McGinnis K, Seshadri V, Dorweiler JE, White J, Sikkink K, Chandler VL (2006) An RNA-dependent RNA polymerase is required for paramutation in maize. Nature 442:295–298
88. Erhard KF Jr, Stonaker JL, Parkinson SE, Lim JP, Hale CJ, Hollick JB (2009) RNA polymerase IV functions in paramutation in *Zea mays*. Science 323:1201–1205
89. Stam M, Belele C, Dorweiler JE, Chandler VL (2002) Differential chromatin structure within a tandem array 100 kb upstream of the maize b1 locus is associated with paramutation. Genes Dev 16:1906–1918
90. Matzke M, Kanno T, Huettel B, Daxinger L, Matzke AJM (2006) RNA-directed DNA methylation and pol IVb in Arabidopsis. Cold Spring Harb Symp Quant Biol 71:449–459
91. Chandler VL, Eggleston WB, Dorweiler JE (2000) Paramutation in maize. Plant Mol Biol 43:121–145
92. Suter CM, Martin DIK (2010) Paramutation: the tip of an epigenetic iceberg? Trends Genet 26:9–14
93. Hollick JB (2010) Paramutation and development. Annu Rev Cell Dev Biol 26:557–579
94. Rassoulzadegan M, Grandjean V, Gounon P, Vincent S, Gillot I, Cuzin F (2006) RNA-mediated non-Mendelian inheritance of an epigenetic change in the mouse. Nature 441:469–474
95. Chandler VL (2007) Paramutation: from maize to mice. Cell 128:641–645
96. Cuzin F, Grandjean V, Rassoulzadegan M (2008) Inherited variation at the epigenetic level: paramutation from the plant to the mouse. Curr Opin Genet Dev 18:193–196
97. de Vanssay A, Bougé AL, Boivin A, Hermant C, Teysset L, Delmarre V, Antoniewski C, Ronsseray S (2012) Paramutation in drosophila linked to emergence of a piRNA-producing locus. Nature 490:112–115
98. Cuzin F, Rassoulzadegan M (2010) Non-Mendelian epigenetic heredity: gametic RNAs as epigenetic regulators and transgenerational signals. Essays Biochem 48:101–106
99. Parkinson SE, Gross SM, Hollick JB (2007) Maize sex determination and abaxial leaf fates are canalized by a factor that maintains repressed epigenetic states. Dev Biol 308:462–473
100. Garnier O, Laoueille-Duprat S, Spillane C (2008) Genomic imprinting in plants. Epigenetics 3:14–20
101. Bauer MJ, Fischer RL (2011) Genome demethylation and imprinting in the endosperm. Curr Opin Genet Dev 14:162–167
102. Goday C, Esteban MR (2001) Chromosome elimination in sciarid flies. Bioessays 23: 242–250
103. Kermicle JL (1970) Dependence of the R-mottled aleurone phenotype in maize on mode of sexual transmission. Genetics 66:69
104. McGrath J, Solter D (1984) Completion of mouse embryogenesis requires both the maternal and paternal genomes. Cell 37:179–183
105. Surani MAH, Barton SC, Norris ML (1984) Development of reconstituted mouse eggs suggests imprinting of the genome during gametogenesis. Nature 308:548–550
106. Hirasawa R, Feil R (2010) Genomic imprinting and human disease. Essays Biochem 48: 187–200
107. Köhler C, Wolff P, Spillane C (2012) Epigenetic mechanisms underlying genomic imprinting in plants. Annu Rev Plant Biol 63:331–352
108. Bauer MJ, Fischer RL (2011) Genome demethylation and imprinting in the endosperm. Curr Opin Plant Biol 14:162–167
109. Raissig MT, Baroux C, Grossniklaus U (2011) Regulation and flexibility of genomic imprinting during seed development. Plant Cell 23: 16–26
110. Grossniklaus U, Vielle-Calzada J-P, Hoeppner MA, Gagliano WB (1998) Maternal control of embryogenesis by *MEDEA*, a Polycomb group gene in Arabidopsis. Science 280: 446–450
111. Josefsson C, Dilkes B, Comai L (2006) Parent-dependent loss of gene silencing during interspecies hybridization. Curr Biol 16:1322–1328

112. Luo M, Taylor JM, Spriggs A, Zhang H, Wu X, Russell S, Singh M, Koltunow A (2011) A genome-wide survey of imprinted genes in rice seeds reveals imprinting primarily occurs in the endosperm. PLoS Genet 7:e1002125
113. Waters AJ, Makarevitch I, Eichten SR, Swanson-Wagner RA, Yeh C-T, Xu W, Schnable PS, Vaughn MW, Gehring M, Springer NM (2011) Parent-of-origin effects on gene expression and DNA methylation in the maize endosperm. Plant Cell 23:4221–4233
114. McKeown PC, Laouielle-Duprat S, Prins P, Wolff P, Schmid MW, Donoghue MT, Fort A, Duszynska D, Comte A, Lao NT, Wennblom TJ, Smant G, Kohler C, Grossniklaus U, Spillane C (2011) Identification of imprinted genes subject to parent-of-origin specific expression in *Arabidopsis thaliana* seeds. BMC Plant Biol 11:113
115. Hsieh T-F, Shin J, Uzawa R, Silva P, Cohen S, Bauer MJ, Hashimoto M, Kirkbride RC, Harada JJ, Zilberman D, Fischer RL (2011) Regulation of imprinted gene expression in Arabidopsis endosperm. Proc Natl Acad Sci U S A 108:1755–1762
116. Wolff P, Weinhofer I, Seguin J, Roszak P, Beisel C, Donoghue MTA, Spillane C, Nordborg M, Rehmsmeier M, Köhler C (2011) High-resolution analysis of parent-of-origin allelic expression in the Arabidopsis endosperm. PLoS Genet 7:e1002126
117. Moore T, Haig D (1991) Genomic imprinting in mammalian development a parental tug-of-war. Trends Genet 7:45–49
118. Tucker S, Vitins A, Pikaard CS (2010) Nucleolar dominance and ribosomal RNA gene silencing. Curr Opin Cell Biol 22:351–356
119. McClintock B (1934) The relation of a particular chromosomal element to the development of the nucleoli in *Zea mays*. Cell Tissue Res 21:294–326
120. Pikaard CS (2000) The epigenetics of nucleolar dominance. Trends Genet 16:495–500
121. Honjo T, Reeder RH (1973) Preferential transcription of *Xenopus laevis* ribosomal RNA in interspecies hybrids between *Xenopus laevis* and *Xenopus mulleri*. J Mol Biol 80:217–228
122. McStay B (2006) Nucleolar dominance: a model for rRNA gene silencing. Genes Dev 20:1207–1214
123. McStay B, Grummt I (2008) The epigenetics of rRNA genes: from molecular to chromosome biology. Annu Rev Cell Dev Biol 24: 131–157
124. Flavell RB, Odell M, Thompson WF (1988) Regulation of cytosine methylation in ribosomal RNA and nucleolus organizer expression in wheat. J Mol Biol 204:523–534
125. Chen ZJ, Pikaard CS (1997) Epigenetic silencing of RNA polymerase I transcription: a role for DNA methylation and histone modification in nucleolar dominance. Genes Dev 11:2124–2136
126. Conconi A, Widmer RM, Koller T, Sogo JM (1989) Two different chromatin structures coexist in ribosomal RNA genes throughout the cell cycle. Cell 57:753–761
127. Pontes O, Lawrence RJ, Neves N, Silva M, Lee JH, Chen ZJ, Viegas W, Pikaard CS (2003) Natural variation in nucleolar dominance reveals the relationship between nucleolus organizer chromatin topology and rRNA gene transcription in Arabidopsis. Proc Natl Acad Sci U S A 100:11418–11423
128. Liu X, Yu C-W, Duan J, Luo M, Wang K, Tian G, Cui Y, Wu K (2012) HDA6 directly interacts with DNA methyltransferase MET1 and maintains transposable element silencing in Arabidopsis. Plant Physiol 158:119–129
129. Lawrence RJ, Earley K, Pontes O, Silva M, Chen ZJ, Neves N, Viegas W, Pikaard CS (2004) A concerted DNA methylation/histone methylation switch regulates rRNA gene dosage control and nucleolar dominance. Mol Cell 13:599–609
130. McKeown PC, Shaw P (2009) Chromatin: linking structure and function in the nucleolus. Chromosoma 118:11–23
131. Pontes O, Li CF, Nunes PC, Haag J, Ream T, Vitins A, Jacobsen SE, Pikaard CS (2006) The Arabidopsis chromatin-modifying nuclear siRNA pathway involves a nucleolar RNA processing center. Cell 126:79–92
132. Kim D-H, Doyle MR, Sung S, Amasino RM (2009) Vernalization: winter and the timing of flowering in plants. Annu Rev Cell Dev Biol 25:277–299
133. Andrés F, Coupland G (2012) The genetic basis of flowering responses to seasonal cues. Nat Rev Genet 13:627–639
134. Whyte RO, Hudson PS (1933) Vernalization or Lyssenko's method for the pre-treatment of seed. Imp Bur Plant Genet 27:1
135. Amasino R (2004) Vernalization, competence, and the epigenetic memory of winter. Plant Cell 16:2553–2559
136. Chouard P (1960) Vernalization and its relations to dormancy. Annu Rev Plant Physiol 11:191–238
137. Caspari EW, Marshak RE (1965) Rise and fall of Lysenko. Science 149:275–278
138. Sung SB, Amasino RM (2004) Vernalization and epigenetics: how plants remember winter. Curr Opin Plant Biol 7:4–10
139. He Y (2009) Control of the transition to flowering by chromatin modifications. Mol Plant 2:554–564

140. Ahmad A, Zhang Y, Cao X-F (2010) Decoding the epigenetic language of plant development. Mol Plant 3:719–728
141. Sung SB, Amasino RM (2004) Vernalization in *Arabidopsis thaliana* is mediated by the PHD finger protein VIN3. Nature 427: 159–164
142. Swiezewski S, Liu F, Magusin A, Dean C (2009) Cold-induced silencing by long antisense transcripts of an Arabidopsis Polycomb target. Nature 462:799–802
143. Sicard A, Lenhard M (2011) The selfing syndrome: a model for studying the genetic and evolutionary basis of morphological adaptation in plants. Ann Bot 107:1433–1443
144. Winn AA, Elle E, Kalisz S, Cheptou P-O, Eckert CG, Goodwillie C, Johnston MO, Moeller DA, Ree RH, Sargent RD, Vallejo-Marín M (2011) Analysis of inbreeding depression in mixed-mating plants provides evidence for selective interferences and stable mixed mating. Evolution 65:3339–3359
145. Pennisi E (2011) Epigenetics linked to inbreeding depression. Science 333:1563
146. Biemont C (2010) Inbreeding effects in the epigenetic era. Nat Rev Genet 11:234
147. Springer NM, Stupar RM (2007) Allelic variation and heterosis in maize: how do two halves make more than a whole? Genome Res 17:264–275
148. Duszynska D, McKeown PC, Juenger TE, Pietraszewska-Bogiel A, Geelen D, Spillane C (2013) Gamete fertility and ovule number variation in selfed reciprocal F1 hybrid triploid plants are heritable and display epigenetic parent-of-origin effects. New Phytol 198: 71–81
149. Yao H, Gray AD, Auger DL, Birchler JA (2013) Genomic dosage effects on heterosis in triploid maize. Proc Natl Acad Sci U S A 110:2665–2669
150. Kaplan Z, Fehrer J (2007) Molecular evidence for a natural primary triple hybrid in plants revealed from direct sequencing. Ann Bot 99:1213–1222
151. Crow JF (1999) Anecdotal, historical and critical commentaries on genetics. Genetics 152:821–825
152. He G, Zhu X, Elling AA, Chen L, Wang X, Guo L, Liang M, He H, Zhang H, Chen F, Qi Y, Chen R, Deng X-W (2010) Global epigenetic and transcriptional trends among two rice subspecies and their reciprocal hybrids. Plant Cell 22:17–33
153. Chodavarapu RK, Feng S, Ding B, Simon SA, Lopez D, Jia Y, Wang GL, Meyers BC, Jacobsen SE, Pellegrini M (2012) Transcriptome and methylome interactions in rice hybrids. Proc Natl Acad Sci U S A 109:12040–12045
154. Groszmann M, Greaves IK, Albertyn ZI, Scofield GN, Peacock WJ, Dennis ES (2011) Changes in 24-nt siRNA levels in Arabidopsis hybrids suggest an epigenetic contribution to hybrid vigor. Proc Natl Acad Sci U S A 108:2617–2622
155. Comai L (2000) Genetic and epigenetic interactions in allopolyploid plants. Plant Mol Biol 43:387–399
156. Gaeta RT, Pires JC (2010) Homoeologous recombination in allopolyploids: the polyploid ratchet. New Phytol 186:18–28
157. Liu B, Wendel JF (2003) Epigenetic phenomena and the evolution of plant allopolyploids. Mol Phylogenet Evol 29:365–379
158. Parisod C, Holderegger R, Brochmann C (2010) Evolutionary consequences of autopolyploidy. New Phytol 186:5–17
159. Paun O, Forest F, Fay MF, Chase MW (2009) Hybrid speciation in angiosperms: parental divergence drives ploidy. New Phytol 182:507–518
160. Soltis PS, Soltis DE (2009) The role of hybridization in plant speciation. Annu Rev Plant Biol 60:561–588
161. Hegarty MJ, Hiscock SJ (2008) Genomic clues to the evolutionary success of review polyploid plants. Curr Biol 18:R435–R444
162. Donoghue MTA, Fort A, Clifton R, Zhang X, McKeown PC, Voigt-Zielinski ML, Borevitz JO, Spillane C (2013) CmCGG methylation-independent parent-of-origin effects on genome-wide transcript levels in isogenic reciprocal F1 triploid plants. DNA Research doi:10.1093/dnares/dst046
163. McKeown PC, Fort A, Duszynska D, Sulpice R, Spillane C. (2013) Emerging molecular mechanisms for biotechnological harnessing of heterosis in crops. Trends Biotechnol 31:549–551
164. Birchler JA, Yao H, Chudalayandi S (2006) Unraveling the genetic basis of hybrid vigor. Proc Natl Acad Sci U S A 103:12957–12958
165. Groszmann M, Greaves IK, Albert N, Fujimoto R, Helliwell CA, Dennis ES, Peacock WJ (2011) Epigenetics in plants—vernalisation and hybrid vigour. Biochim Biophys Acta 1809:427–437
166. Thudi M, Li Y, Jackson SA, May GD, Varshney RK (2012) Current state-of-art of sequencing technologies for plant genomics research. Brief Funct Genomics 11:3–11
167. Schmitz RJ, Zhang X (2011) High-throughput approaches for plant epigenomic studies. Curr Opin Plant Biol 14:130–136
168. ENCODE Project Consortium (2012) An integrated encyclopedia of DNA elements in the human genome. Nature 489:57–74
169. Hawkins RD, Hon GC, Ren B (2010) Next-generation genomics: an integrative approach. Nat Rev Genet 11:476–486
170. Shen W-H, Xu L (2009) Chromatin remodeling in stem cell maintenance in *Arabidopsis thaliana*. Mol Plant 2:600–609

# Chapter 2

# The Gene Balance Hypothesis: Dosage Effects in Plants

James A. Birchler and Reiner A. Veitia

## Abstract

The concept of genomic balance traces to the early days of genetics. In recent years, studies of gene expression have found parallels to the classical phenotypic studies in that aneuploid changes have greater effects than whole genome changes. This has an explanation in terms of potential stoichiometric imbalances of the gene products encoded in the aneuploid regions. Studies of transcriptional factor mutations indicated that they tend to be haplo-insufficient as heterozygotes. Molecular evolution studies found that genes encoding members of macromolecular complexes were preferentially retained following polyploidy and underrepresented in copy number variants. In this review chapter, we synthesize these observations under the rubric of the Gene Balance Hypothesis.

**Key words** Aneuploidy, Ploidy, Copy number variants, Quantitative traits, Gene expression, Dosage compensation

## 1 Introduction to the Gene Balance Hypothesis

The Gene Balance Hypothesis posits that varying the stoichiometry of members of multi-subunit complexes will affect the function of the whole complex as a result of the topology, kinetics, and mode of assembly [1–5]. This principle applies to any type of macromolecular complex but perhaps its most critical implications are in the area of gene regulation, which is mediated in large part by oligomeric complexes. Because varying the stoichiometry of subunits has an effect, this will be manifested in a dosage response when the encoding gene is varied in copy number. Thus, gene regulatory systems tend to be dosage-dependent and thus will impact quantitative characteristics. The general idea of balance traces to the early days of genetics [6–8] but more recently a synthesis pulling together data from quantitative genetics, biophysics, molecular evolution, and studies of gene expression has been formulated. In this review article, we summarize the evidence for this synthesis and note some implications.

Charles Spillane and Peter C. McKeown (eds.), *Plant Epigenetics and Epigenomics: Methods and Protocols*,
Methods in Molecular Biology, vol. 1112, DOI 10.1007/978-1-62703-773-0_2, 

One of the lines of evidence for the dosage balance concept is the classical observation that aneuploidy is generally more severe than ploidy changes. This concept was first formulated by Blakeslee and colleagues using the flowering plant, Datura [6, 7]. Trisomics were isolated for each of the 12 chromosomes. Each exhibited a characteristic phenotype. In comparison, a whole genome ploidy series was produced by chromosome doubling. The phenotypic changes in this case were not as dramatic as for the individual chromosome copy number modulations. This relationship has been found in many other plant and animal species over the subsequent decades [9, 10].

More recently, studies of gene expression modulation in aneuploidy and ploidy series showed that a greater number of modulations were found with aneuploidy than ploidy-level variations, in parallel to the phenotypic effects. There are two major types of modulations in aneuploids. One involves positive correlations with the varied chromosome that act in *trans* across the genome. The other type of *trans*-acting effect found was an inverse correlation of gene expression with the dosage of the varied chromosome [12]. These effects were found on the enzyme activity [11], protein [12], and messenger RNA levels [13]. In the latter study, the modulations caused by chromosomal dosage were within the direct and inverse correlative levels in the diploid embryo as well as in the triploid endosperm. In other words, the magnitude of genomic imbalance at the respective ploidy levels determined the magnitude of the effects. Changes of whole ploidy show fewer effects [14].

For the genes on the varied chromosome, it is generally assumed that a structural gene dosage effect will be produced with a change of chromosomal dosage. This is indeed the case for many gene products, but many cases of dosage compensation were also observed [11, 15, 16]. Dosage compensation is the phenomenon that the same amount of gene product is produced regardless of the chromosomal dosage. Examples of the *alcohol dehydrogenase 1* [11] and the *PRO* [12] genes located on the long arm of chromosome 1 exhibited the same amount of gene product in a 1-to-3 dosage series of this chromosome arm. In the case of *adh1*, the basis of the compensation was shown to be that an inverse dosage effect was operating on the locus in question which counteracted the structural gene dosage effect that might otherwise occur [12, 16]. Division of the long arm of chromosome 1 revealed a region that produces an inverse dosage effect upon *adh1* and that varying the dosage of a small region around *adh1* itself produced a gene dosage effect [16].

## 2 Gene Balance and Aneuploidy

The basis of the aneuploid effect was shown to be able to be reduced to the action of single genes [17, 18]. The leaky *white-apricot* allele of the *white* eye color gene in Drosophila was used as a reporter to identify modifiers that would increase or decrease the amount of pigment when the new mutation was heterozygous. This situation would mimic a "monosomic" condition but on the single gene level. From over 2 decades of screening, 47 such modifiers were identified [18]. The majority of them acted negatively. Such a large number of modifiers are likely to result from the fact that many processes operate through regulatory hierarchies and/or through oligomeric regulatory factors. Each modifier would affect overlapping sets of genes.

This type of result has parallels in the genetics of quantitative traits. Quantitative trait loci are usually additive and multigenic [19] as are aneuploid syndromes [13]. Furthermore, they are controlled by many genes usually of small effect that are additive [20–25]. In other words, there is a dosage effect of the controlling alleles. Thus, there are similarities among the control of quantitative traits, the impact of multiple aneuploidies on the phenotype and the multigenic set of modifiers identified for a single phenotype [18].

Indeed, of the quantitative trait loci whose molecular nature has been elucidated, they are typically some type of regulatory factor. The first QTL cloned and molecularly characterized was *fw2.2*, which controls fruit weight in tomato [20]. When a transgenic dosage series was produced for this gene, a negative dosage effect on fruit weight was realized [26]. Among the collection of modifiers of the *white* eye color gene, those whose molecular nature is known consist of transcription factors, signal transduction components, and chromatin-modifying factors [18].

Another line of evidence in support of the Gene Balance Hypothesis is that haplo-insufficient genes in yeast and humans are enriched for proteins within complexes [27–30]. While these genes include the spectrum of those involved in macromolecular complexes, they include transcription factors and signal transduction components. The concept of balance was examined by over-expression of the same genes, which was found to be detrimental also [29]. However, co-over-expression was capable of correcting the fitness defects of interactors [29].

Further evidence comes from studies of molecular evolution. Throughout the plant kingdom [31–37], but also in yeast [38] and the animal kingdom [39, 40], there have been cycles of whole genome duplication (polyploidization) following by fractionation (diploidization). As genes are lost in the latter process, there is not a random distribution of the functional classes of genes that are retained [34–36]. Indeed, there is a preferential retention of genes whose products are involved with macromolecular complexes [34–36]. Included among these are transcription factors and signal transduction components. The implication is that if the stoichiometry of these gene products is important, deletion of one member of a duplicate pair might act like an aneuploid effect and be selected against, thus resulting in retention over longer periods of evolutionary time than other classes of genes.

The reciprocal result is found for segmental duplications and copy number variants. In this case there is an underrepresentation of genes whose products are involved in oligomeric complexes [34–36, 41–44]. Instead, genes encoding products that provide a selective advantage via greater quantity without balance defects are preferentially represented in partial genome duplications. This principle is reinforced by the realization that proteins that are increasingly under-wrapped (a measure of the reliance of a protein on binding partnerships to maintain structural integrity) are less likely to be correlated with gene duplicability [45]. Indeed, an inverse relationship between the extent of protein under-wrapping and gene family size has been demonstrated. Thus, gene duplication is unlikely to be tolerated if the structure of the corresponding protein requires substantial protein–protein stabilizing interactions unless the latter are co-duplicated or co-retained. Moreover, copy number polymorphisms in Drosophila [43] and humans [46] for genes with network centrality are significantly underrepresented.

Lastly, there are constraints on the tolerated variation of regulatory genes. In *Paramecium tetraurelia*, which has experienced three detectable whole genome duplication events as revealed by the genome sequence, there is evidence of purifying selection, based on Ka/Ks ratios, on the coding sequence of both members of a retained duplicate pair implying that dosage is important [39]. Because the conserved duplicate genes are likely to have kept the ancestral function, neofunctionalization cannot explain their retention. Instead, this result might be explained if mutations that upset the stoichiometric balance are selected against leaving the sequence signature of purifying selection. A similar conclusion can be drawn from an illuminating mutation accumulation experiment in *C. elegans* [47]. Mutations were allowed to accumulate and then patterns of gene expression were measured. Considerable variation for changes in the expression of individual target genes was revealed but there was conservation of the global patterns of gene expression suggesting that purifying selection was occurring for changes in the quantities of regulatory factors [47].

In a similar vein, studies of *cis* and *trans* variation in gene expression in general find that *cis* variation is typically of greater magnitude, although less pleiotropic, than *trans* variation but for any one modulation of a gene product there is a greater number of these more subtle changes [48–60]. This type of result would occur if target genes were not constrained for the type of *cis* regulatory variation that could be tolerated (probably within limits) but that the multiple regulatory genes have a constraint on the magnitude of variation that can be tolerated and maintained in populations.

## 3 Implications of the Gene Balance Hypothesis

The Gene Balance Hypothesis suggests that new mutations in regulatory genes of various types will likely produce a semidominant dosage effect to some degree and to have a (subtle) phenotypic effect. The consequence of this is that new mutations will be available for selection, be that either purifying or adaptive. Mutations that are completely recessive are not available for selection. They may become lost in a population or alternatively, only in a small population would drift and inbreeding make them homozygous and thus responsive to selective forces. The implication is that there is a greater availability for adaptive selection for regulatory genes than for others that do not exhibit dosage stoichiometries.

While new dosage-sensitive mutations would be readily available for selection, it is likely that this property of regulatory genes would also work to maintain the status quo in regulatory processes due to purifying selection against detrimental mutations perturbing the stoichiometric balance. It is generally considered that purifying selection is more common than adaptive selection, but once adaptations occur, purifying selection would maintain them.

Another principle suggested by the results described above is that regulatory changes would have an impact on evolution in subtle increments but that many genes can contribute to any one trajectory. The evidence, noted above, from the study of modifiers of the *white* eye color gene and from quantitative trait multigenic control, illustrates that many genes can impact a single phenotypic characteristic. The data from retention of duplicate genes encoding macromolecular complexes following ancient polyploidization events and their underrepresentation in copy number variants suggests that the magnitude of tolerable dosage effect is narrow and well below a twofold range. Thus, the standing variation in regulatory processes is likely to be quite subtle but would be contributed by many genes. Thus, the control of quantitative traits will be determined by many genes each with a small effect.

Future studies involved with the Gene Balance Hypothesis might focus on the effect of stoichiometric changes of individual subunits of macromolecular complexes and how these changes alter the function of the whole complex. Some possibilities might be that the kinetics of assembly lead to unproductive partial complexes [3] or that targeted degradation of unused subunits may alleviate or, on the contrary, enhance dosage effects [5]. Another question involves how new balances are achieved during evolution. As noted above, *cis* variation will accumulate in target genes and eventually will be in conflict with the *trans* regulatory system if critical target genes change their expression. The evolutionary evidence from preferential retention following polyploidization suggests that there is resistance to altered balance but ultimately this would change and elucidating the processes by which this occurs would be illuminating. Further, microRNAs are known to impact gene expression in a dosage-sensitive manner and so they are likely to play a role in gene balance mechanisms but basically nothing is known about this possibility at present. Lastly, it is of interest to decipher whether issues of regulatory gene balance play any role in speciation [3]. If new balances are indeed achieved in separate evolutionary lineages, then their combination in hybrids might prevent gene flow by causing reduced fitness at some level.

## Acknowledgements

Research in our labs is supported by National Institutes of Health grant RO1GM068042-05 and National Science Foundation grant DBI 0733857 Plant Genome.

### References

1. Birchler JA, Riddle NC, Auger DL, Veitia RA (2005) Dosage balance in gene regulation: biological implications. Trends Genet 21: 219–226
2. Birchler JA, Veitia RA (2007) The gene balance hypothesis: from classical genetics to modern genomics. Plant Cell 19:395–402
3. Birchler JA, Veitia RA (2010) The gene balance hypothesis: implications for gene regulation, quantitative traits and evolution. New Phytol 186:54–62
4. Birchler JA, Yao H, Chudalayandi S (2007) Biological consequences of dosage dependent gene regulatory systems. Biochim Biophys Acta 1769:422–428
5. Veitia RA, Bottani S, Birchler JA (2008) Cellular reactions to gene dosage imbalance: genomic, transcriptomic and proteomic effects. Trends Genet 24:390–397
6. Blakeslee AF, Belling J, Farnham ME (1920) Chromosomal duplication and Mendelian phenomena in Datura mutants. Science 52: 388–390
7. Blakeslee AF (1934) New Jimson weeds from old chromosomes. J Hered 24:80–108
8. Bridges CB (1925) Sex in relation to chromosomes and genes. Am Nat 59:127–137
9. Lindsley DL, Sandler L, Baker BS, Carpenter AT, Denell RE, Hall JC, Jacobs PA, Miklos GL, Davis BK, Gethman RC et al (1972) Segmental aneuploidy and the genetic gross structure of the Drosophila genome. Genetics 71:157–184
10. Bond DJ, Chandley AC (1983) Aneuploidy. Oxford University Press, Oxford
11. Birchler JA (1979) A study of enzyme activities in a dosage series of the long arm of chromosome one in maize. Genetics 92:1211–1229
12. Birchler JA, Newton KJ (1981) Modulation of protein levels in chromosomal dosage series of maize: the biochemical basis of aneuploid syndromes. Genetics 99:247–266
13. Guo M, Birchler JA (1994) Trans-acting dosage effects on the expression of model gene systems in maize aneuploids. Science 266:1999–2002
14. Guo M, Davis D, Birchler JA (1996) Dosage effects on gene expression in a maize ploidy series. Genetics 142:1349–1355
15. Birchler JA, Hiebert JC, Paigen K (1990) Analysis of autosomal dosage compensation involving the alcohol dehydrogenase locus in *Drosophila melanogaster*. Genetics 124:677–686
16. Birchler JA (1981) The genetic basis of dosage compensation of alcohol dehydrogenase-1 in maize. Genetics 97:625–637
17. Rabinow L, Nguyen-Huynh AT, Birchler JA (1991) A trans-acting regulatory gene that inversely affects the expression of the white, brown and scarlet loci in *Drosophila melanogaster*. Genetics 129:463–480
18. Birchler JA, Bhadra U, Pal Bhadra M, Auger DL (2001) Dosage dependent gene regulation in multicellular eukaryotes: implications for dosage compensation, aneuploid syndromes and quantitative traits. Dev Biol 234:275–288
19. Tanksley SD (1993) Mapping polygenes. Annu Rev Genet 27:205–233
20. Frary A, Nesbitt RC, Frary A, Grandillo S, van der Knaap E, Cong B, Liu J, Meller J, Elber R,

Alpert KB, Tanksley SD (2000) *fw2.2*: a quantitative trait locus key to the evolution of tomato fruit size. Science 289:85–88
21. Cong B, Liu J, Tanksley SD (2002) Natural alleles at a tomato fruit size quantitative trait locus differ by heterochronic regulatory mutations. Proc Natl Acad Sci U S A 99: 13606–13611
22. Cong B, Barrero LS, Tanksley SD (2008) Regulatory change in YABBY-like transcription factor led to evolution of extreme fruit size during tomato domestication. Nat Genet 40:800–804
23. Liu J, Van Eck J, Cong B, Tanksley SD (2002) A new class of regulatory genes underlying the cause of pear-shaped tomato fruit. Proc Natl Acad Sci U S A 99:13302–13306
24. Burke JM, Tang S, Knapp SJ, Rieseberg LH (2002) Genetic analysis of sunflower domestication. Genetics 161:1257–1267
25. Buckler ES, Holland JB, Bradbury PJ, Acharya CB, Brown PJ et al (2009) The genetic architecture of maize flowering time. Science 325:714–718
26. Liu J, Cong B, Tanksley SD (2003) Generation and analysis of an artificial gene dosage series in tomato to study the mechanisms by which the cloned quantitative trait locus *fw2.2* controls fruit size. Plant Physiol 132:292–299
27. Veitia RA (2002) Exploring the etiology of haploinsufficiency. Bioessays 24:175–184
28. Seidman JG, Seidman C (2002) Transcription factor haploinsufficiency: when half a loaf is not enough. J Clin Invest 109:451–455
29. Papp B, Pal C, Hurst LD (2003) Dosage sensitivity and the evolution of gene families in yeast. Nature 424:194–197
30. Kondrashov FA, Koonin EV (2004) A common framework for understanding the origin of genetic dominance and evolutionary fates of gene duplications. Trends Genet 20:287–290
31. Simillion C, Vandepoele K, Montagu MC, Zabeau M, Van de Peer Y (2002) The hidden duplication past of *Arabidopsis thaliana*. Proc Natl Acad Sci U S A 99:13627–13632
32. Bowers JE, Chapman BA, Rong J, Paterson AH (2003) Unraveling angiosperm genome evolution by phylogenetic analysis of chromosomal duplication events. Nature 422: 433–438
33. Chapman BA, Bowers JE, Feltus FA, Paterson AH (2006) Buffering of crucial functions by paleologous duplicated genes may contribute cyclicality to angiosperm genome duplication. Proc Natl Acad Sci U S A 103:2730–2735
34. Maere S, DeBodt S, Raes J, Casneuf T, Van Montagu M, Kuiper M, Van de Peer Y (2005) Modeling gene and genome duplications in eukaryotes. Proc Natl Acad Sci U S A 102: 5454–5459
35. Blanc G, Wolfe KH (2004) Functional divergence of duplicated genes formed by polyploidy during Arabidopsis evolution. Plant Cell 16:1679–1691
36. Freeling M, Thomas BC (2006) Gene-balanced duplications, like tetraploidy, provide predictable drive to increase morphological complexity. Genome Res 16:805–814
37. Barker MS, Kane NC, Matvienko M, Kozik A, Michelmore RW, Knapp SJ, Rieseberg LH (2008) Multiple paleopolyploidizations during the evolution of the compositae reveal parallel patterns of duplicate gene retention after millions of years. Mol Biol Evol 25:2445–2455
38. Wolfe KH, Shields DC (1997) Molecular evidence for an ancient duplication of the entire yeast genome. Nature 387:708–713
39. Aury J-M, Jaillon O, Duret L, Jubin C, Porcel BM, Segurens B, Daubin V, Anthouard V, Aiach N, Arnaiz O et al (2006) Global trends of whole-genome duplications revealed by the ciliate *Paramecium tetraurelia*. Nature 444: 171–178
40. Blomme T, Vandepoele K, De Bodt S, Simillion C, Maere S, Van de Peer Y (2006) The gain and loss of genes during 600 million years of vertebrate evolution. Genome Biol 7:R43
41. Freeling M, Lyons E, Pedersen B, Alam M, Ming R, Lisch D (2008) Many or most genes in Arabidopsis transposed after the origin of the order Brassicales. Genome Res 18:1924–1937
42. Davis JC, Petrov DA (2005) Do disparate mechanisms of duplication add similar genes to the genome? Trends Genet 21:548–551
43. Dopman EB, Hartl DL (2007) A portrait of copy-number polymorphism in *Drosophila melanogaster*. Proc Natl Acad Sci U S A 104: 19920–19925
44. Hakes L, Pinney JW, Lovell SC, Oliver SG, Robertson DL (2007) All duplicates are not equal: the difference between small-scale and genome duplications. Genome Biol 8:R209
45. Liang H, Rogale-Plazonic K, Chen J, Li WH, Fernandez A (2008) Protein under-wrapping causes dosage sensitivity and decreases gene duplicability. PLoS Genet 4:e11
46. Schuster-Bockler B, Conrad D, Bateman A (2010) Dosage sensitivity shapes the evolution of copy-number varied regions. PLoS One 5:e9474
47. Denver DR, Morris K, Streelman JT, Kim SK, Lynch M, Thomas WK (2005) The transcriptional consequences of mutation and natural selection in *Caenorhabditis elegans*. Nat Genet 37:544–548
48. Schadt EE, Monks SA, Drake TA, Lusis AJ, Che N, Colinayo V, Ruff TG, Milligan SB, Lamb JR, Cavet G, Linsley PS, Mao M, Stoughton RB, Friend SH (2003) Genetics of gene expression surveyed in maize, mouse and man. Nature 422:297–302

49. Yvert G, Brem RB, Whittle J, Akey JM, Foss E, Smith EN, Mackelprang R, Kruglyak L (2003) Trans-acting regulatory variation in *Saccharomyces cerevisiae* and the role of transcripton factors. Nat Genet 35:57–64
50. Wittkopp PJ, Haerum BK, Clark AG (2004) Evolutionary changes in cis and trans gene regulation. Nature 430:85–88
51. Wayne ML, Pan Y-J, Nuzhdin SV, McIntyre LM (2004) Additivity and trans-acting effects on gene expression in male *Drosophila simulans*. Genetics 168:1413–1420
52. Morley M, Molony CM, Weber TM, Devlin JL, Ewens KG, Spielman RS, Cheung VG (2004) Genetic analysis of genome-wide variation in human gene expression. Nature 430:743–747
53. Hughes KA, Ayroles JF, Reedy MM, Drnevich JM, Rowe KC, Ruedi EA, Caceres CE, Paige KN (2006) Segregating variation in the transcriptome: cis regulation and additivity of effects. Genetics 173:1347–1355
54. West MAL, Kim K, Kliebenstein DJ, van Leeuwen H, Michelmore RW, Doerge RW, St. Clair DA (2007) Global eQTL mapping reveals the complex genetic architecture of transcript-level variation in Arabidopsis. Genetics 175:1441–1450
55. Wang D, Sung H-M, Wang T-Y, Huang C-J, Yang P, Chang T, Wang Y-C, Tseng D-L, Wu J-P, Lee T-C, Shih M-C, Li W-H (2007) Expression evolution in yeast genes of single-input modules is mainly due to changes in trans-acting factors. Genome Res 17:1161–1169
56. Petretto E, Mangion J, Dickens NJ, Cook SSA, Kumaran MK, Lu M, Fischer J, Maatz H, Kren V, Pravenec M, Hubner N, Aitman TJ (2006) Heritability and tissue specificity of expression quantitative trait loci. PLoS Genet 2:e172
57. Grieve IC, Dickens NJ, Pravenec M, Kren V, Hubner N, Cook SA, Ailtman TJ, Petretto E, Mangion J (2008) Genome-wide co-expression analysis in multiple tissues. PLoS One 3:e4033
58. Lemos B, Araripe LO, Fontanilla P, Hartl DL (2008) Dominance and the evolutionary accumulation of cis- and trans-effects on gene expression. Proc Natl Acad Sci U S A 105:1813–1822
59. L'Hote D, Serres C, Veitia RA, Montagutelli X, Oulmouden A, Vaiman D (2008) Gene expression regulation in the context of mouse interspecific mosaic genomes. Genome Biol 9:R133
60. Tirosh I, Reikhav S, Levy AA, Barkai N (2009) A yeast hybrid provides insight into the evolution of gene expression regulation. Science 324:659–662

# Chapter 3

# High-Throughput RNA-Seq for Allelic or Locus-Specific Expression Analysis in *Arabidopsis*-Related Species, Hybrids, and Allotetraploids

**Danny W-K. Ng, Xiaoli Shi, Gyoungju Nah, and Z. Jeffrey Chen**

## Abstract

With the next generation sequencing technology, RNA-Seq (RNA sequencing) becomes one of the most powerful tools in quantification of global transcriptomes, discovery of new transcripts and alternative isoforms, as well as detection of single nucleotide polymorphisms (SNPs). RNA-Seq is advantageous over hybridization-based gene quantification methods: (1) it does not require prior information about genomic sequences, (2) it avoids high background problem caused by cross-hybridization, and (3) it is highly sensitive and avoids background and saturation of signals; and finally it is capable of detecting allelic expression differences in hybrids and allopolyploids. We used the RNA-Seq method to determine the genome-wide transcriptome changes in Arabidopsis allotetraploids and their parents, *A. thaliana* and *A. arenosa*. The use of this approach allows us to quantify transcriptome from these species and more importantly, to identify allelic or homoeologous-specific gene expression that plays a role in morphological evolution of allopolyploids. The computational pipelines developed are also applicable to the analysis of chromatin immunoprecipitation sequencing (ChIP-seq) data in Arabidopsis-related species, hybrids, and allopolyploids. Comparative analysis of RNA-Seq and ChIP-Seq data will allow us to determine the effects of chromatin modifications on nonadditive gene expression in hybrids and allopolyploids.

**Key words** RNA-Seq, Next generation sequencing, Transcriptome, Read mapping, Arabidopsis, Allopolyploids, Allelic expression

## 1 Introduction

Genome-wide gene expression changes have been documented in allopolyploids of many species including Arabidopsis [1, 2], cotton [3], Senecio [4], and wheat [5, 6]. Arabidopsis is a model system for the study of gene expression changes in response to auto- and allopolyploidization [7–9]. Using oligo-gene microarrays, Wang et al. [1] found that 15–43 % of genes are differentially expressed between the two closely related Arabidopsis species, *A. thaliana* (At4; $2n=4x=20$) and *A. arenosa* (Aa; $2n=4x=32$) [1]. In the allotetraploids (Allos; $2n=4x=26$), 5–38 % of genes are expressed

Charles Spillane and Peter C. McKeown (eds.), *Plant Epigenetics and Epigenomics: Methods and Protocols*, Methods in Molecular Biology, vol. 1112, DOI 10.1007/978-1-62703-773-0_3, 

nonadditively (different from the mid-parent value) relative to the progenitors. Nonadditive gene expression is associated with non-additive phenotypes such as large stature, growth vigor, and late flowering observed in the allotetraploids [1, 2, 10], many of which are controlled by genetic and epigenetic mechanisms [8]. Nonadditively expressed genes are enriched in energy, metabolism, stress response, and phytohormonal regulation [1, 11]. Further analysis has demonstrated that the growth vigor in Arabidopsis allotetraploids and hybrids is linked directly with epigenetic regulation of circadian regulators that control downstream pathways in chlorophyll biosynthesis and starch metabolism [12].

Although hybridization-based high-throughput methods, like microarrays, provide an enormous amount of gene expression information, the technology is limited by the availability of probe sequence information and their specificity in hybridization [13, 14]. In the species whose genomes are not sequenced, there are computational errors associated with probe design and EST annotation. A major drawback of the array technology is its incapability of discriminating between paralogues or homologues that have similar sequences such as homoeologous loci in allopolyploids that are derived from related progenitors. RNA sequencing (RNA-Seq) or mRNA-Seq refers to the use of high-throughput sequencing technologies to sequence cDNAs in order to get a complete inventory of RNAs in a given sample [15, 16]. To study genome-wide differential allelic expression in allopolyploids, we used the RNA-Seq method for transcriptome analysis in *A. thaliana*, *A. arenosa*, and their allotetraploids in $F_1$ and $F_7$ generations. This analysis allowed us to investigate homoeologous gene expression changes contributed by the two diverged genomes within the allotetraploids.

## 2 Materials

### 2.1 Plant Materials and Growth Media

1. Plant lines are available from the Arabidopsis Biological Resource Center (ABRC) including *Arabidopsis thaliana* autotetraploid (At4; ABRC, CS3900), *A. arenosa* (Aa; ABRC, CS3901), resynthesized allotetraploids in $F_1$ and $F_7$ (Allo733; ABRC, CS3895) generations from At4 and Aa crosses.
2. The growth media include Murashige and Skoog basal medium powder with sucrose and agar (MS agar) and plant tissue culture tested (Sigma-Aldrich, St. Louis, MO). Growth media are prepared by dissolving 42.4 g MS agar in a final volume of 1 l with water (*see* **Note 1**). The pH of media is adjusted to 5.8 with 1 M potassium hydroxide solution (KOH). Sterilization of media is performed at 121 °C for 20 min at 15 psi.

### 2.2 RNA Isolation

1. Plant RNA reagent (Invitrogen, Carlsbad, CA).
2. DEPC-treated water: Prepare by mixing 0.1 % DEPC (Diethylpyrocarbonate; v/v) with water and incubate for 1 h at

37 °C. Autoclave the DEPC-treated water at 121 °C for 30 min at 15 psi to inactivate DEPC before use.

3. Chloroform, certified ACS (Fisher Scientific, Waltham, MA).
4. Isopropanol (Fisher Scientific, Waltham, MA).
5. Nuclease-free water (Ambion Inc., Foster City, CA).
6. Sodium Chloride (NaCl): Prepare 5 M NaCl stock solution with DEPC-treated water.

### 2.3 Denaturing Agarose Gel Electrophoresis of RNA

1. GenePure LE quick dissolve agarose (Bioexpress, Kaysville, UT).
2. 4-morpholino propane sulfonic acid (MOPS) running buffer (10×): 0.2 M MOPS (molecular biology grade, Sigma-Aldrich), 50 mM sodium acetate-$3H_2O$, 10 mM ethylene diamine tetraacetic acid disodium salt (EDTA-$Na_2$), dissolve in DEPC-treated water, and adjust to pH 7 with 10 M sodium hydroxide (*see* **Note 2**).
3. Formaldehyde solution, 37 % (Sigma-Aldrich) (*see* **Note 3**).
4. NorthernMax Formaldehyde loading dye (Ambion Inc.).
5. Ethidium bromide (Fisher Scientific).
6. RNaseZap® RNase decontamination solution (Ambion Inc.).

### 2.4 RNA-Seq Library Preparation

1. RNA-Seq 8-Sample Prep Kit (Illumina Inc., San Diego, CA).
2. Maximum recovery microcentrifuge tubes (1.7 ml) (Axygen, Union City, CA).
3. DynaMag-2 Magnetic Particle Concentrator (Invitrogen).
4. Superscript II reverse transcriptase with 100 mM DTT and 5× first strand buffer (Invitrogen).
5. DNA marker: 1 kb Plus DNA ladder (Invitrogen).
6. DNA purification: MinElute PCR purification kit, QIAquick gel extraction kit, QIAquick PCR purification kit (Qiagen, Germantown, MD).
7. Certified low-range Ultra agarose (BioRad, Hercules, CA).
8. TAE electrophoresis buffer (1×): 40 mM Tris-base, 5.71 ml glacial acetic acid, 1 mM EDTA. Store at room temperature.
9. Sodium acetate: Prepare 3 M sodium acetate and adjust pH to 5.2 with glacial acetic acid.
10. DNA loading dye (6×): 40 % sucrose (w/v), 0.25 % bromophenol blue, and 0.25 % xylene cyanol FF.
11. GeneCatcher disposable gel excision kit (Gel Company, San Francisco, CA).
12. Dark reader transilluminators (Clare Chemical Research, Dolores, CO).

## 3 Methods

The RNA-Seq approach is divided into three major processes: (1) RNA-Seq sample preparation/sequencing, (2) mapping raw reads, and (3) quantification of gene expression. Uniformity of sequence coverage is an important issue because it can affect sensitivity in detection, accuracy in quantification, and complete connection in exon-intergenic regions [16, 17]. Hydrolysis of RNA samples before the cDNA synthesis step dramatically improved the uniformity of sequence coverage because cDNA priming at putative random sites may be biased towards enrichment of 5′-ends of transcripts [16, 18]. During mapping of raw reads to the reference sequence, the results often fall into either unique-mapped reads or multi-mapped reads. Although unique-reads themselves are often used for transcript quantification, multi-reads can be included for accurate measurement. Transcript level is defined in reads per kilo base of exon model per million mapped reads (RPKM) [16, 17]. The RPKM value is used for direct comparison of transcript levels between samples.

### 3.1 Preparing Plant Materials

1. Sterilize Arabidopsis seeds in 1 ml 100 % Clorox in a 1.7 ml tube for 3 min with shaking. Spin down seeds in a microcentrifuge to remove Clorox (5 s) and wash seeds with 1 ml water for 3 min with shaking. Repeat the washing step for five times before plating seeds onto MS agar media.
2. Keep the plated seeds in a cold room (4 °C) for 48 h before transferring to a growth chamber with setting at 16/8 h (light/dark) cycles at 22 °C. Transfer seedlings to soil for further growth at 22 °C with 16/8 h (light/dark) cycles after 2–3 weeks.
3. Collect fresh mature leaves prior to bolting (6–8 rosette leaves from 3 to 4 weeks old *A. thaliana*, 6–7 weeks old *A. arenosa* or allotetraploid plants) for RNA isolation.

### 3.2 Total RNA Isolation

1. Grind mature leaves into fine powder under liquid nitrogen using mortar and pestle.
2. Transfer 100–150 mg tissues into a 1.7 ml tube.
3. Add 500 μl Plant RNA reagent (Invitrogen) to frozen ground tissue. Homogenize samples by vortexing and lay the tube down horizontally (to maximize surface area for RNA extraction) for 5 min at room temperature (r.t.).
4. Centrifuge samples at 12,000 × *g* for 2 min at r.t. and transfer 500 μl supernatant to a new 1.7 ml tube containing 100 μl 5 M NaCl and mix.
5. Add 300 μl Chloroform and mix thoroughly.
6. Separate the aqueous phase by centrifuge at 12,000 × *g* for 10 min at 4 °C and transfer the aqueous phase (~550 μl) into a new 1.7 ml tube.

7. Add an equal volume of isopropanol, mix, and incubate the sample at r.t. for 10 min.
8. Centrifuge at 12,000 ×*g* for 10 min at 4 °C to pellet RNA and then remove supernatant.
9. Wash pellet with 1 ml 75 % ethanol and centrifuge at 12,000 ×*g* for 1 min at r.t.
10. Remove supernatant and air dry pellet (~3–5 min).
11. Resuspend RNA in 40 μl nuclease-free water (Ambion).
12. Take 1 μl RNA for quantitation using NanoDrop spectrophotometers (Fisher Scientific). The yield is typically around 60–120 μg.
13. Check the integrity and quality of RNA by gel electrophoresis (Subheading 3.3) or store RNA at −80 °C.

### 3.3 Formaldehyde Denaturing Agarose Gel Electrophoresis for Total RNA Quality Validation

The following protocol is based on the use of FisherBiotech horizontal electrophoresis system with a 7 × 10 cm (W × L) gel size. It is important that the gel apparatus (including gel tray and gel tank) are wiped with RNaseZap® solution to remove RNase contamination and rinsed with DEPC-treated $H_2O$. All solutions should be prepared using DEPC-treated $H_2O$.

1. Prepare 1 % agarose gel by heating 0.9 g GenePure LE quick dissolve agarose in 78.5 ml DEPC-treated $H_2O$ until dissolve.
2. Cool down gel to 70 °C and add 9 ml 10× MOPS buffer and 37 % formaldehyde solution (*see* **Note 4**).
3. Pour agarose gel into the gel tray and let it to solidify at r.t. for 1–2 h.
4. Make up 1 μg RNA samples to 3 μl with nuclease-free $H_2O$ (Ambion) and add 3 volumes of NorthernMax formaldehyde loading dye.
5. Denature RNA at 65 °C in a water bath for 15 min to remove RNA secondary structure.
6. Place the samples on ice and add ethidium bromide or a reduced-toxicity alternative to a final concentration of 50 μg/ml.
7. Load the samples on the denaturing formaldehyde agarose gel and perform electrophoresis at 80 V for 1 h and 45 min in 1× MOPS buffer until the bromophenol blue dye front has migrated 2/3 the length of the gel.
8. Visualize gel with a UV transilluminator to determine the integrity of the prepared RNA samples. Good quality and intact RNA samples will have sharp 28S and 18S ribosomal RNA bands (Fig. 1). Degraded RNA will appear as a smear in the gel.

### 3.4 RNA-Seq Library Preparation

For RNA-Seq library preparation, majority of the reagents used are provided in the RNA-Seq 8-sample prep kit (Illumina Inc.). Additional reagents are listed under Subheading 2.4. Since particular

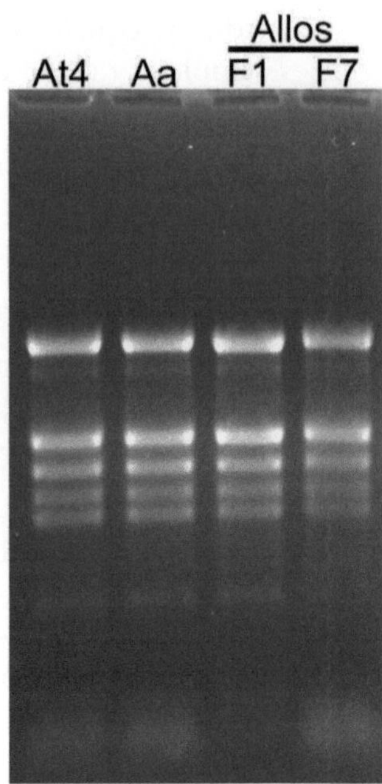


**Fig. 1** Verification of RNA quality in a denaturing agarose gel. Total RNA (1 μg) from mature leaves of various plant lines is resolved on a 1 % formaldehyde denaturing agarose gel. At4, *Arabidopsis thaliana* autotetraploid; Aa, *Arabidopsis arenosa* autotetraploid; Allos, $F_1$ or $F_7$ generation of allotetraploids between At4 and Aa

details of the protocol could change from time to time, the full protocol is not described here. A detail protocol is provided by the supplier with the kit (www.illumina.com). However, several critical steps within the protocol are listed below. The entire process typically takes 2 days to accomplish and can be divided into six major steps including:

1. Purify poly-A containing mRNA from total RNA using magnetic beads with attached poly-T oligos (*see* **Note 5**).
2. Use high temperature to fragment the purified mRNA.
3. Convert the mRNA fragments into first cDNA strand by reverse transcription with random primers and synthesized the second strand cDNA using DNA polymerase I (DNA pol I).
4. Repair cDNA ends using a series of reactions involving T4 DNA polymerase, Klenow DNA polymerase, and adapter oligos ligation following phosphorylation of 5′ end of cDNA with T4 polynucleotide kinase and A-tailing with Klenow fragment (3′–5′ exo minus).
5. Purify a selected size range of cDNA templates from agarose gel for subsequent PCR amplification of library (*see* **Note 6**).
6. Validate the RNA-Seq library by resolving 10 % purified cDNA library in a 2 % agarose gel (*see* **Note** 7).

### 3.5 Mapping of Sequence Reads

Once sequencing data were generated using Illumina genome analyzer or HighSeq 2000, the sequencing coverage was estimated based on total gene number, average gene size in the organism, as well as read length and total read number. Raw sequencing reads were processed by first mapping them using BFAST, a Blast-like Fast Accurate Search Tool (http://bfast.sourceforge.net) against the Arabidopsis reference genome (ftp://ftp.arabidopsis.org/home/tair/Sequences/whole_chromosomes/) and a cDNA (ftp://ftp.arabidopsis.org/home/tair/Sequences/blast_datasets/TAIR9_blastsets/TAIR9_cdna_20090619) database. Because most existing aligners for short reads are not able to map the reads that span introns, alignments with two databases are used to identify the reads that mapped to single exon and those that span introns within a gene model separately. To assign multi-reads originating from duplicated regions, all mapped reads are separated into unique and multiple reads. A preliminary measure of expression calculated by allocating unique reads to duplicate gene models can be used to weight the multiple reads mapped to multiple loci. For instance, if a read mapped to two duplicates with equal scoring and the ratio of the preliminary measure of their expression is 2:1, the multiple read contributes 2/3 and 1/3 to two duplicates individually [16]. Finally, all mapped reads are identified and categorized into four classes: unique best reads, multiple best reads, unique splicing reads, and multiple splicing reads. Classification of reads is implemented by a C++ package developed in house (Shi, unpublished).

1. Calculate sequence coverage by dividing product of read length and total read number by total length of exon regions in the whole genome.
2. Generate index files for two reference databases, genome and cDNA, by complementing BFAST index (http://bfast.sourceforge.net).
3. Split raw sequencing reads files into several smaller files by implementing Unix split command (*see* **Note 8**).
4. Use the indexed genome sequences database from **step 2** to map the raw read sequences.
5. Categorize mapped reads into two classes based on the number of matches having the best score. One class contains only unique best scoring alignments—unique best matches, and the other contains multiple matches having the best score—multiple best matches.
6. Filter out all unique and multiple best matches from raw read data files and map the remaining reads to the indexed cDNA sequences database from **step 2** to identify exon-splicing matches.
7. Categorize mapped reads spanning introns into two classes based on the number of matches having the best score. One contains unique splicing matches and the other contains multiple splicing matches.

### 3.6 Quantify Transcript Levels in Reads per Kilobase of Exon Model per Million Mapped Reads

In quantifying gene expression, normalization is an important issue in comparing expression level between and within species. To normalize the read quantity for transcript level estimation, RPKM [16] is used to take into consideration transcript length and the total number of mapped reads affecting read quantity. This approach is able to distinguish transcript levels for isoforms and duplicates. The expression levels of isoforms are identified by allocating unambiguous reads, which are unrelated to any other isoforms, to calculate preliminary RPKM, and then weighting ambiguous reads using the preliminary RPKM to quantify transcript levels for isoforms. The expression levels of duplicates are identified by assigning unique reads to calculate preliminary RPKM and then weighting multiple reads using the preliminary RPKM to quantify transcript levels for duplicates (*see* introduction to Subheading 3.5). A simplified version of peak search method encapsulated in ELAND [16] is used to identify novel transcripts covered by highly abundant reads. The quantification of the transcriptome is computed by a C++ package developed in house (Shi, unpublished).

1. Assign unique best reads and unique splicing reads to known gene models or segments unrelated with any isoforms to calculate preliminary RPKM [16].
2. Identify novel exons by sliding a 100 bp window and aggregating mapped reads within window outside known exons along the genome.
3. Merge connected windows containing at least four 60 bp length reads (*see* **Note 9**). Boundaries of a newly identified exon are the farthest points of a peak whose coverage is at least 0.1-fold of the peak coverage (*see* **Note 10**).
4. Identify novel transcripts by combining neighboring novel exons with comparable expression levels.
5. Calculate RPKM for all known and newly identified transcripts by allocating unique best reads and unique splicing reads. Unique reads associated with isoforms are weighted by preliminary RPKM of corresponding transcripts computed at **step 1**.
6. Calculate final RPKM for all known and newly identified transcripts by allocating all mapped reads. Multiple reads coming from particular locus are weighted by RPKM of associated transcripts computed at **step 5**.

### 3.7 Allele-Specific Expression Estimation in Allotetraploid Hybrids

A critical point for estimating allelic expression in allotetraploids is to identify parental genotypes of the reads in the RNA-Seq library. In the absence of the complete genome sequence of *A. arenosa*, a single nucleotide polymorphism (SNP) database is created using sequence reads from leaves, siliques, and normalized tissues in *A. arenosa*. The three libraries generated a highly consistent set of SNPs, and over 97 % of SNPs are identical. At the genome-wide level, an average of 27 SNPs was detected per kilobase sequence of exon. The SNPs were distributed through 23,041 genes. This SNP database is then used to assign reads from the allotetraploids into genotype-specific reads from either the *A. thaliana* or the *A. arenosa* genome. This will allow the calculation of allelic expression patterns of homoeologous loci in the allotetraploids.

#### *3.7.1 SNP Discovery*

1. Map sequence reads from two parents to the available reference genome (TAIR9 for *A. thaliana* ftp://ftp.arabidopsis.org/home/tair/Sequences/whole_chromosomes/).
2. Assemble mapped reads based on the coordinates of the reference genome.
3. Compare the assembled mRNAs from the two parents and choose the differential nucleotides at the same coordinate of the reference sequence.
4. Filter out SNPs with low read coverage empirically in terms of accuracy required for a given experiment (*see* **Note 11**).
5. Use the created SNP database to assign reads as in Subheading 3.7.2.

#### *3.7.2 Mapping Allelic Reads from Allotetraploids and Allelic Expression Estimation*

1. Map reads from allotetraploids to the reference genome and categorize mapped reads to four classes (Subheading 3.5, **steps 2–7**).
2. Quantify the transcript level (RPKM) for each locus (Subheading 3.6, **steps 1–6**).
3. Assign the mapped reads from allotetraploids (Subheading 3.5, **step** 7) to their corresponding parental genotype using the SNP database (Subheading 3.7.1).
4. Using the ratio of assigned mapped reads, estimate the frequency of allelic expression contributed by homoeologous loci from two genomes in allotetraploids.
5. To evaluate the accuracy of the estimation, proceed to Subheading 3.8.1.
6. If no correction is required for the allelic expression frequency estimation, proceed to **step** 7.
7. Estimate genotype-specific allele expression level by multiplying the allelic expression frequency and the overall expression level of both homoeologous alleles determined in (Subheading 3.6) (*see* **Note 12**).

### *3.8 Correction for Allele-Specific Expression Frequency Estimation*

In allotetraploids, mapping reads could be biased because of sequence divergence between homoeologous alleles, degree of heterozygosity, and without *A. arenosa* reference genome. Reads originating from alleles with close relationship to the reference genome are more likely to be matched. In addition, the sequence polymorphism between two parental alleles among individuals within population could cause biased sensitivity when SNPs were used to identify reads that originated from different genotypes. Therefore it is necessary to evaluate the accuracy of allelic expression frequency estimation and correct the potential bias.

#### *3.8.1 Detection of Systematic Errors for Allele-Specific Expression Frequency Estimation*

An artificial read mixture with an equal number of known reads coming from each of the two parents is used to assess the accuracy of allelic expression estimation.

1. Select equally quantitative reads derived from assembled mRNA of the two progenitors of allotetraploids obtained from **step 2** under Subheading 3.7.1.
2. Mix the selected reads to generate a simulated reads data set with 50 % expected frequency of allelic expression.
3. Map simulated reads to the reference genome and categorize them to the four classes following **steps 2–7** under the Subheading 3.5.
4. Estimate frequency of allelic expression following **steps 2–4** of Subheading 3.7.2.
5. Compare the expected and calculated allelic expression frequency by performing a paired *t*-test.
6. If significant differences exist between the expected and calculated allelic frequency, proceed to Subheading 3.8.2 to correct the derived allele expression frequency.

#### *3.8.2 Correction of Systematic Errors of Allele-Specific Expression Frequency Estimation*

If systematic errors exist in estimating the allele-specific expression frequency in allotetraploids, a statistical approach integrating simulation and regression analysis can be used to derive a function between the expected and estimated frequencies for each gene, thus permitting correction of the systematic errors.

1. Randomly select reads from assembled mRNAs originating from the two progenitors of the allotetraploids obtained from **step 2** under Subheading 3.7.1.
2. Construct three mixed-read groups according to three expected ratios of allelic expression, 1:3, 1:1 (or 2:2), and 3:1, for each assembled transcript.
3. Estimate allelic expression frequency using these three simulated datasets following **steps 1–4** in Subheading 3.7.2.
4. Derive the function between the expected and estimated frequencies using a linear regression analysis.
5. Correct the estimated allelic expression frequency using the function derived from the regression analysis.
6. Use the corrected estimated allelic expression frequency to calculate the genotype-specific allele expression level (Subheading 3.7.2, **step 7**) (Fig. 2).

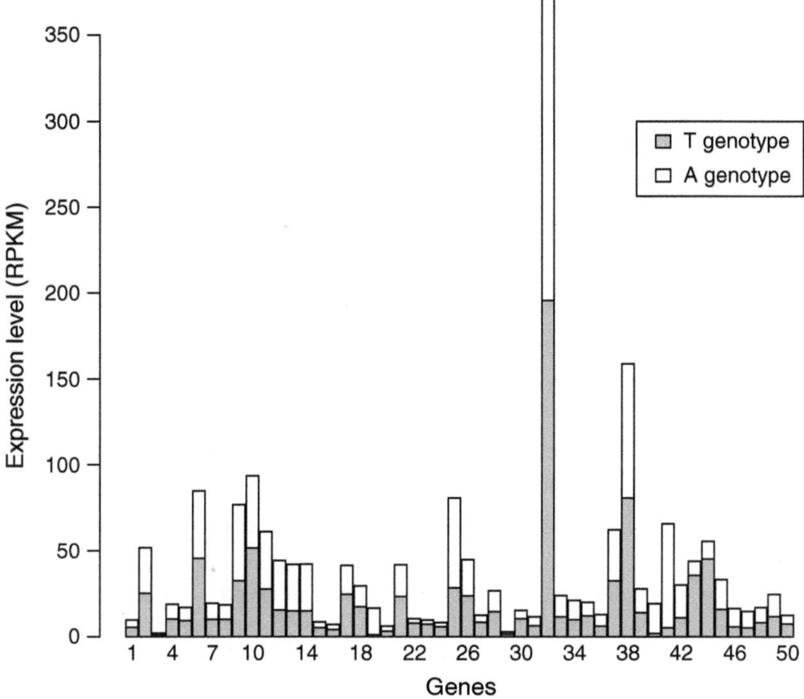


**Fig. 2** A histogram showing gene- and allele-specific expression of 50 loci on a chromosome of resynthesized allotetraploid ($F_1$). The height of bar is RPKM value of each locus. *Grey* and *white* indicate digital expression levels of *A. thaliana* (T) and *A. arensoa* (A) genotypes, respectively

### 3.9 Statistical Analysis of RNA-Seq and Microarray Data

Expression microarrays have been the main method for transcriptome analysis since the mid-1990s [15]. Recent developments of ultra-high-throughput mRNA sequencing techniques provide a simpler and more comprehensive way to measure transcriptome composition [16]. The methods of transcriptome measurement are different in RNA-Seq and microarray data processing. Raw microarray data have to be processed through several steps, including background-correction, normalization, computation of expression values based on probe intensities [19, 20]. After the data are normalized and the quantification is completed, similar statistical tests can be applied to analyze expression values for both RNA-Seq and microarray data. Some recent publications have described Fisher's exact test and likelihood ratio test for identifying differentially expressed genes from the RNA-Seq data [15, 21]. A newly developed R package DEGseq represents a novel method based on the MA-plot to identify differentially expressed genes [22]. The MA-plot-based method is able to handle models with or without technical replicates. The latter includes some additional steps to estimate the noise level of genes with different expression levels. An assessment of technical reproducibility of RNA-Seq and gene expression arrays [15] suggests that RNA-Seq experiments typically have low background noise and that their data fit the Poisson model well [22]. Although biological replications are important for statistical analysis,

RNA-Seq data used without replication will also generate useful expression information [15]. The p-values derived from multiple statistical tests should be corrected for occurrence of false positives by multiple testing corrections, such as two types of correction provided in the DEGseq [22].

## 4 Notes

1. Unless specified, all water used has a resistivity of 18.2 MΩ cm. To sterilize water or solutions, autoclave them at 121 °C for 30 min at 15 psi. For procedures involving RNA handling, treat water with 0.1 % DEPC (Diethylpyrocarbonate; v/v) and incubate for at least 1 h at 37 °C. To inactivate DEPC before use, autoclave DEPC-treated water at 121 °C for 30 min at 15 psi.
2. MOPS buffer is light- and temperature-sensitive, store it at room temperature and protect it from light. Do not use the buffer if it becomes dark in color (oxidized). When autoclaved, the MOPS buffer will become yellowish indicating degradation. We recommend filter sterilization of the buffer.
3. Formaldehyde is a probable human carcinogen and toxic by inhalation, it should be handled inside a chemical fume hood. If the chemical appears cloudy with precipitates, this indicates degradation and it should be disposed following the guidelines of your institution/organization.
4. Formaldehyde is a volatile organic compound and its vapor is toxic and can cause irritation. It is highly recommended to prepare the denaturing formaldehyde agarose gel in a chemical fume hood, especially when adding formaldehyde into the warm agarose gel solution.
5. The amount of total RNA used can range from 1 to 10 μg. Unless specified, use maximum recovery 1.7 ml microcentrifuge tubes to maximize sample recovery when preparing the RNA-Seq library. When purifying mRNA using magnetic beads, allow enough time for the beads to be captured by the magnetic stand. Avoid drying the magnetic beads when exchanging buffers. After the beads have been resuspended by vortexing, perform a quick spin (<1 s) with a benchtop minicentrifuge to bring all beads to the bottom of the tube (not pellet the beads) for efficient beads capture with the magnetic stand. The eluted mRNA after the first bead binding and bead washing steps are subjected to a second oligo-dT bead binding and bead washing, this is to maximize the purity of mRNA for library generation.
6. In visualizing gels for size selection of cDNA templates, the use of a Dark reader is preferable. The Dark reader transilluminator

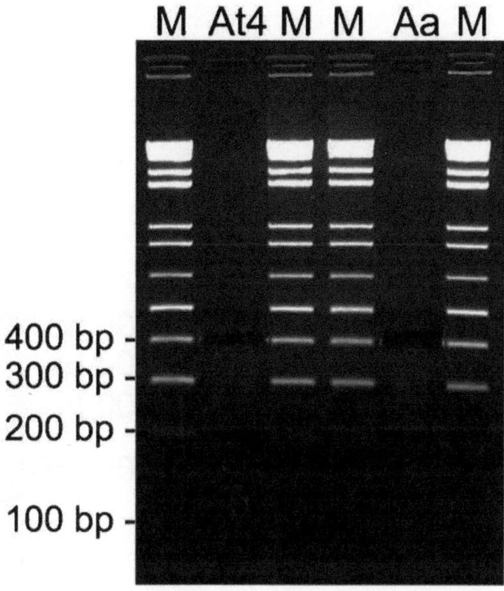


**Fig. 3** Purification of cDNA templates. Representative ethidium bromide stained cDNA templates from *A. thaliana* (At4) and *A. arenosa* (Aa) on a 2 % agarose gel (certified low-range ultra agarose). Templates at 200 bp (±25 bp) and 400 bp (±25 bp) size ranges are excised from the gel. M, 1 kb + DNA ladder (Invitrogen)

uses visible blue light as the excitation source; therefore, it reduces UV-mediated DNA damages and is safer to operate when compared to an UV transilluminator. However, a UV transilluminator could be used for excising gel by keeping the UV exposure of DNA minimal. With improving sequencing cluster generation chemistry, we recommend excising gel at a 400 bp (±25 bp) size range (Fig. 3) instead of the 200 bp (±25 bp) range suggested in the supplier manual. It is normal that cDNA is not visible in the gel. If an UV transilluminator is used for gel excision, including fewer samples in one gel as this also helps to minimize the DNA's exposure to UV.

7. The cDNA library should be visible as a distinct band at approximately the selected size (Fig. 4). As an alternative to agarose gel electrophoresis, the quality of the prepared library can be analyzed using a 2100 Bioanalyzer (Agilent Technologies). In addition, clone 1 μl amplified library into a sequencing vector (with blunt ends) and sequence clones with a conventional DNA sequencing approach.
8. To facilitate multitask running on multiple processors, split reads files into several smaller files before read mapping.
9. The threshold of read counts is derived by assuming that the read count of a gene satisfies the Poisson distribution. The mean and

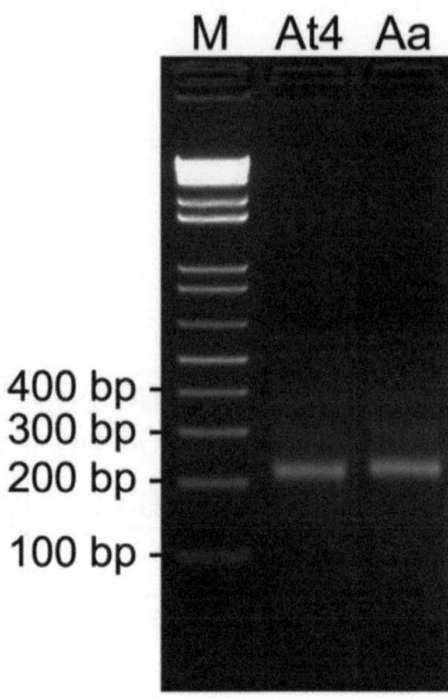


**Fig. 4** Quantity and quality of size-selected cDNA. Amplified cDNAs of 200 bp (±25 bp) from *A. thaliana* (At4) and *A. arenosa* (Aa) were resolved in a 2 % agarose gel and stained with ethidium bromide. M, 1 kb + DNA ladder (Invitrogen)

variance of the Poisson distribution are calculated from the read counts within a 100 bp length gene expressed at the median level of RPKMs for all known gene models. The threshold of read counts is the lowest value within a 95 % confidence interval of the Poisson distribution.

10. The fold change of read coverage between boundaries and peaks could be adjusted empirically in terms of distribution of exon length across the whole genome.
11. To obtain more reliable SNPs, low quality SNPs could be filtered out based on numbers of reads supporting each SNP because SNPs supported by few reads might have high variation between population of a species.
12. Further improvement of the estimation of allelic expression frequency can be achieved using a statistical model to infer the frequencies of allele expression via an expectation maximization (EM) algorithm [23, 24].

Although our analysis is focused on RNA-Seq data, the computational pipelines developed are applicable to analysis of ChIP-seq data obtained from sequencing the DNA prepared from chromatin immunoprecipitation (ChIP) [25] (http://www.natureprotocols.com/2009/01/08/chromatin_immunoprecipitation_2.php) in related species, and hybrids and allopolyploids.

## Acknowledgements

We thank Luca Comai and his lab members for sharing *A. arenosa* SNP databases for mapping the reads. The work was supported by grants from the Plant Genome Research Program of the National Science Foundation (DBI0733857 to Z.J.C.).

## References

1. Wang J, Tian L, Lee HS, Wei NE, Jiang H, Watson B, Madlung A, Osborn TC, Doerge RW, Comai L, Chen ZJ (2006) Genomewide nonadditive gene regulation in *Arabidopsis* allotetraploids. Genetics 172:507–517
2. Wang J, Tian L, Madlung A, Lee HS, Chen M, Lee JJ, Watson B, Kagochi T, Comai L, Chen ZJ (2004) Stochastic and epigenetic changes of gene expression in Arabidopsis polyploids. Genetics 167:1961–1973
3. Flagel L, Udall J, Nettleton D, Wendel J (2008) Duplicate gene expression in allopolyploid Gossypium reveals two temporally distinct phases of expression evolution. BMC Biol 6:16
4. Hegarty MJ, Barker GL, Wilson ID, Abbott RJ, Edwards KJ, Hiscock SJ (2006) Transcriptome shock after interspecific hybridization in senecio is ameliorated by genome duplication. Curr Biol 16:1652–1659
5. Pumphrey M, Bai J, Laudencia-Chingcuanco D, Anderson O, Gill BS (2009) Nonadditive expression of homoeologous genes is established upon polyploidization in hexaploid wheat. Genetics 181:1147–1157
6. Chague V, Just J, Mestiri I, Balzergue S, Tanguy AM, Huneau C, Huteau V, Belcram H, Coriton O, Jahier J, Chalhoub B (2010) Genome-wide gene expression changes in genetically stable synthetic and natural wheat allohexaploids. New Phytol 187:1181–1194
7. Osborn TC, Pires JC, Birchler JA, Auger DL, Chen ZJ, Lee HS, Comai L, Madlung A, Doerge RW, Colot V, Martienssen RA (2003) Understanding mechanisms of novel gene expression in polyploids. Trends Genet 19:141–147
8. Chen ZJ (2007) Genetic and epigenetic mechanisms for gene expression and phenotypic variation in plant polyploids. Annu Rev Plant Biol 58:377–406
9. Chen ZJ, Ni Z (2006) Mechanisms of genomic rearrangements and gene expression changes in plant polyploids. Bioessays 28:240–252
10. Comai L, Tyagi AP, Winter K, Holmes-Davis R, Reynolds SH, Stevens Y, Byers B (2000) Phenotypic instability and rapid gene silencing in newly formed *Arabidopsis* allotetraploids. Plant Cell 12:1551–1568
11. Jackson S, Chen ZJ (2010) Genomic and expression plasticity of polyploidy. Curr Opin Plant Biol 13:153–159
12. Ni Z, Kim ED, Ha M, Lackey E, Liu J, Zhang Y, Sun Q, Chen ZJ (2009) Altered circadian rhythms regulate growth vigour in hybrids and allopolyploids. Nature 457:327–331
13. Hoheisel JD (2006) Microarray technology: beyond transcript profiling and genotype analysis. Nat Rev Genet 7:200–210
14. Sobek J, Bartscherer K, Jacob A, Hoheisel JD, Angenendt P (2006) Microarray technology as a universal tool for high-throughput analysis of biological systems. Comb Chem High Throughput Screen 9:365–380
15. Marioni JC, Mason CE, Mane SM, Stephens M, Gilad Y (2008) RNA-seq: an assessment of technical reproducibility and comparison with gene expression arrays. Genome Res 18: 1509–1517
16. Mortazavi A, Williams BA, McCue K, Schaeffer L, Wold B (2008) Mapping and quantifying mammalian transcriptomes by RNA-Seq. Nat Methods 5:621–628
17. Pepke S, Wold B, Mortazavi A (2009) Computation for ChIP-seq and RNA-seq studies. Nat Methods 6:S22–S32
18. Wang Z, Gerstein M, Snyder M (2009) RNA-Seq: a revolutionary tool for transcriptomics. Nat Rev Genet 10:57–63
19. Gautier L, Cope L, Bolstad BM, Irizarry RA (2004) affy—analysis of Affymetrix GeneChip data at the probe level. Bioinformatics 20:307–315
20. Quackenbush J (2001) Computational analysis of microarray data. Nat Rev Genet 2:418–427
21. Bloom JS, Khan Z, Kruglyak L, Singh M, Caudy AA (2009) Measuring differential gene expression by short read sequencing: quantitative comparison to 2-channel gene expression microarrays. BMC Genomics 10:221

22. Wang L, Feng Z, Wang X, Zhang X (2010) DEGseq: an R package for identifying differentially expressed genes from RNA-seq data. Bioinformatics 26:136–138
23. Eriksson N, Pachter L, Mitsuya Y, Rhee SY, Wang C, Gharizadeh B, Ronaghi M, Shafer RW, Beerenwinkel N (2008) Viral population estimation using pyrosequencing. PLoS Comput Biol 4:e1000074
24. Do CB, Batzoglou S (2008) What is the expectation maximization algorithm? Nat Biotechol 26:897–899
25. Saleh A, Alvarez-Venegas R, Avramova Z (2008) An efficient chromatin immunoprecipitation (ChIP) protocol for studying histone modifications in Arabidopsis plants. Nat Protoc 3:1018–1025

# Chapter 4

# Inference of Allele-Specific Expression from RNA-seq Data

**Paul K. Korir and Cathal Seoighe**

## Abstract

The differential abundance of transcripts from alternative alleles of a gene, for example in a hybrid plant or an outbred natural population, can provide information about the nature of interindividual or interstrain variation in gene expression. Allele-specific expression (ASE) can result from epigenetic phenomena, such as imprinting (when the overexpressed allele is inherited consistently from one parent) or allele-specific chromatin modifications. Alternatively, DNA sequence variants in the promoter or within the transcribed region of a gene can affect the rate of transcription or the rate of decay of the transcript, respectively. The existence of this allelic variation and the insights it provides into the nature of the gene regulation are of significant interest. With the recent widespread availability of sequencing based transcriptomics, the power to detect ASE has increased; however, inference of ASE from transcriptome sequencing data is subject to several caveats and potential biases and the results need to be interpreted with care.

**Key words** Allele-specific expression, RNA-seq, ASE, High-throughput sequencing

## 1 Introduction

Allele-specific expression (ASE), synonymously known as *differential allelic expression* [1–4], *allelic imbalance* (AI) [3, 4] or *allelic bias* [2, 5, 6], is said to occur when parental copies of a gene are expressed at unequal levels. It is identified within an individual by comparing expression levels of alternative alleles at heterozygous loci (not necessarily SNPs). Imprinting is a special case of ASE in which the relative expression level of an allele depends on the parent from which it was inherited as a consequence of differing epigenetic states (e.g., DNA methylation or histone modification) associated with the paternally and maternally inherited alleles. Some ASE experiments have uncovered novel imprinted genes [7]. According to one study, *cis* variants may affect up to 30 % of RefSeq transcripts and over 50 % of *cis* variants give rise to detectable ASE [1]. In plants, one analysis of ASE in maize meristems showed that between 50 and 70 % of genes were differentially expressed between alleles [8].

Charles Spillane and Peter C. McKeown (eds.), *Plant Epigenetics and Epigenomics: Methods and Protocols*,
Methods in Molecular Biology, vol. 1112, DOI 10.1007/978-1-62703-773-0_4, © Springer Science+Business Media New York 2014

### 1.1 Motivation for ASE Inference: Characterization of Cis-Regulatory Variation

Gene expression is a key intermediate, linking genotype with other phenotypes, and the study of gene expression as a quantitative genetic trait has gained in importance in recent years [9]. ASE can provide a means to detect expression quantitative trait loci (eQTLs) and, under certain assumptions, a way to infer the mechanism of action of an eQTL (i.e., *cis*, *trans*, or a combination of the two) [10]. Recently, eQTLs and gene coexpression network analysis have been used to identify genes and gene networks involved in quantitative and complex traits in plants [11], whereas previous studies in *Arabidopsis thaliana* were used to demonstrate the existence of complex expression networks [12].

Variants in *cis* have been shown to have greater effects, on average, on gene expression compared to *trans*-acting variants [12, 13]. Consequently, characterizing all *cis* variants may account for a substantial part of the genetic variation in gene expression. However, in order to detect *cis*-acting expression variants, other sources of gene expression variation (*trans*-acting and environmental), should either be eliminated or strictly controlled. Indeed, part of the goal of ASE inference is to minimize these sources of expression variation by comparing expression levels of alternative alleles within the same sample so that *cis*-acting variants can be detected accurately [2]. In this way, ASE serves as a means to identify *cis*-eQTLs and distinguish them from *trans*-factors. This constitutes an important aspect of the analysis of gene regulation within and between species [5].

Identification of eQTLs allows complex regulatory networks to be dissected, enabling a better understanding of mechanisms controlling gene expression. One genome-wide eQTL analysis in plants made several key observations. First, that transcript-level variation is highly complex even when studied in recombinant inbred lines (RILs), and that expression variation in an outbred population can only be more complex. Second, that a majority of expression traits are influenced by multiple eQTLs (acting both in *cis* and *trans*). Weak individual associations, the majority of which are *trans*-eQTLs, collectively influence gene expression, pointing to complex gene regulation networks [12]. The existence of higher orders of complexity has also been demonstrated, in which several cascades of expression variation are present and may even vary in the course of an organism's lifetime [14].

## 2 Next-Generation Sequencing and RNA-seq

### 2.1 Next-Generation Sequencing

The last decade has seen successive advances in sequencing technologies. Key requirements have been the need to cut costs, increase sequencing throughput and generate sequences of the appropriate quality and length. The achievement of these requirements through the development of several new technologies, often collectively referred to as next-generation sequencing (NGS) technologies, has resulted in the use of sequencing as the method of choice, not only to sequence genomes but also to generate diverse types of genomic data. NGS is now the preferred means to measure gene expression and mRNA splicing (RNA-seq), DNA–protein binding (ChIP-seq, Methyl-seq), DNA accessibility (DNase-seq), metagenomics (pyrosequencing and MPSS) among a range of other applications [15] (as described in Chaps. 6 and 8 of this volume).

### 2.2 RNA and Transcriptomics

The goal of transcriptomics is to shed light on how organisms function at the molecular level by measuring and characterizing the full complement of transcribed RNA [16]. Due to its quantitative nature, RNA-seq has become an important tool in transcriptomics, providing data that may be simultaneously mined for expression profiles, splicing isoforms, and variations in transcription start and end positions. There are several relatively well-established and newer RNA-seq protocols, differing in intermediate purification steps that can have an impact on the quality of reads [17]. By and large, they involve conversion of polyadenylated RNA to double-stranded cDNA followed by adapter ligation, PCR amplification, purification by either hybridization or through the use of padlock probes before, finally, sequencing. Specific subsets of the RNA pool may be targeted, for example by size, or through poly(A) selection steps to increase the representation of mature transcripts. It is important to note that RNA-seq does not imply any particular sequencing platform though the short-read data produced is processed differently for different platforms [15].

### 2.3 Caveats

Meaningful interpretation of generated RNA-seq data requires accurate alignment of sequences, which in turn depend on sequence quality. Therefore, events that occur in the preceding steps that contribute to ambiguous or incorrect alignment can have a severe impact on RNA-seq results. Issues relating to efficiency of mapping algorithms and techniques are covered in [18], but one issue of particular concern for ASE is mapping bias [7]. This is particularly problematic for reads that overlap polymorphic genomic positions. In the case of heterozygotes, reads that have the allele present in the reference sequence are more likely to be mapped correctly. Issues of mapping bias persist even when polymorphic positions are masked by setting them to a nucleotide different to either of the SNP alleles. This has a severe impact on ASE inference for two reasons. First, it means that, because fewer reads end up being aligned at their correct locations, the power to detect ASE is diminished and second, because reads from alternative alleles do not have the same probability of misalignment, false-positive signals of ASE can be obtained [19]. At present, this may be dealt with by first identifying polymorphic positions that have a mapping bias then eliminating them to discard spurious ASE signals. Indeed,

Degner et al. [7] demonstrated that while this eliminated 40 % of the top ASE signals, the remaining positions were enriched for known *cis*-regulatory variants and imprinted loci. RNA-seq is, therefore, a versatile technique with a growing range of applications. Its application to ASE is not without caveats, but it provides a powerful platform through which further insights into gene expression may be obtained.

## 3 Methods

### 3.1 Outline of ASE Inference

We now outline a procedure by which one may carry out inference of ASE from RNA-seq data. The purpose of this chapter is to enable users with little expertise in computer programming or scripting to infer ASE from RNA-seq data obtained from a single individual. We describe the procedure for a Linux environment and provide short Python and R scripts that can be used to process the data generated at each stage of the analysis pipeline—users on other platforms may adapt the steps to suit their computing platform without altering the sequence of steps. In order to illustrate the steps involved, we have chosen some popular and freely available tools. The data is RNA-seq from a Yoruba HapMap cell line [20] accession GM19238 as described in [7].

The directory structure we use is as follows:

- *RNA-seq data sample 1 to N (one folder per sample).* Contains either an SRA file, FASTQ, or FASTA file. SRA files should be converted to FASTQ.
- *SNPs.* Contains data from the HapMap YRI samples to be used for filtering.
- *FASTA reference folder.* Contains the reference genome in FASTA format. This should not be confused with the reference genome index that the mapping application uses.
- *Scripts.* Python and R scripts as detailed below.

We assume that the mapping application and SAMtools are available through the PATH variable.

### 3.2 Obtaining RNA-seq Data

Apart from obtaining data by carrying out sequencing, one readily available source of RNA-seq data is the Sequence Read Archive (SRA), which is a repository for NGS data [21]. Data is hierarchically arranged based on experiments, samples, or runs and indicates key experiment parameters, such as the sequencing platform which was employed. It is important to take note of these as they can have an impact on data analysis by influencing how some steps should be carried out. The native SRA file format has the extension `.sra` and is convertible to FASTQ format using the program `fastq-dump` script from the SRA Toolkit that may be downloaded from the SRA website. All steps that follow will be carried out on the Yoruba lymphoblastoid cell line with HapMap accession GM19238 that was sequenced using Illumina's Genome Analyzer II and consisting of 35 bp reads [7]. This is somewhat shorter than the reads from more recent studies. The proportion of reads that can be mapped uniquely to the reference genome is much higher in the case of longer reads. While there are also some platform-specific considerations, the general principles do not depart from what will be demonstrated.

The lines below are taken from an RNA-seq file with SRA accession number SRR030769. Lines starting with @ are the read identifiers preceding the actual read, whereas those starting with + are the same identifiers preceding read quality scores encoded in ASCII.

```
@SRR030769.1 HWI-EAS279:1:1:0:902 length=35
NACGATCCTTCTGACCTTTTGGGTTTTAAGCAGGA
+SRR030769.1 HWI-EAS279:1:1:0:902 length=35
!/::7:24679:878:::877638:::%%%%%%%%
@SRR030769.2 HWI-EAS279:1:1:0:1765 length=35
NCACAGACGAACACGTGGTGTGCAAAGTCCAGCAC
+SRR030769.2 HWI-EAS279:1:1:0:1765 length=35
!/7888677867666877677666766%%%%%%%%
@SRR030769.3 HWI-EAS279:1:1:0:649 length=35
NTGCAGCGCTGTCTTACCACTGGTGCCCTCCTGCA
+SRR030769.3 HWI-EAS279:1:1:0:649 length=35
!|455157766467866665/4%%%%%%%%%%%%%
```

### 3.3 Mapping

To illustrate the mapping process, we will use Bowtie [22] version 0.12.7 and SAMtools [23] version 0.1.12-10 (r896). Both Bowtie and SAMtools are freely available under Artistic and MIT licenses, respectively. Bowtie requires an index of the reference genome in a compressed .ebwt file that may either be downloaded from the Bowtie website or created using the bowtie-build command and a FASTA file of the organism's genome.

To map reads the following command should be entered:

```
$ bowtie [options] <ebwt> {-1 <m1> -2 <m2> OR
--12 <r> OR <s>} [<hit>]
```

where [options] are optional flags that modify Bowtie's behavior, <ebwt> is the path to the index, {-1... OR <s>} indicates the file containing the reads in one of several structures (*see* Table 1) and [<hit>] indicates an optional file into which the alignment data will be written.

### 3.4 Handling Index Files

Typically, indexes are split into several files that share a basename. The <ebwt> parameter consists of the path and the basename. It might be cumbersome to write the full path on the command-line and environment variables can be employed for this abbreviation. For example:

**Table 1**
**Selected Bowtie parameters**

| Flag | Use |
|---|---|
| `-S` | Output the alignment in SAM format |
| `--solexa-quals`<br>`--solexa1.3-quals` | Converts Solexa input qualities to Phred quality scores. Read quality scores for Illumina (formerly Solexa) are distinguished by the software version of pipeline that is used to create the read files. The former is for pre-v1.3 and the latter for v1.3 to present |
| `-m <int>` | Do not report reads that map to more than `<int>` locations |
| `-n <int>` | Sets one of two runtime parameters: the maximum threshold on the number of mismatches in the first L (specified by `-l` flag—consult Bowtie documentation) nucleotides (the *seed*) OR the maximum threshold on the sum of the Phred quality values at the mismatch positions may not exceed E (specified by `-e` flag); is mutually exclusive with the `-v` flag |
| `-v <int>` | The alignment may have at most `<int>` mismatches |
| `{-1 <m1> -2 <m2>}` | Arguments entered together; each of `m1` and `m2` is one or several files, each with one end of paired-end reads. The order of files in `m1` and `m2` have to correspond to each other |
| `--12 <r>` | A single file of tab-delimited paired-ended reads |
| `<s>` | A file containing single-ended reads |
| `-p <int>` | Sets the number of threads or processors to use |

```
$ export HG19=/<path to index>/hg19
```

for the human genome index hg19 (basename). A more permanent solution would be to include the same command in the user's `.bashrc` file.

The following command outputs alignments in SAM format, excluding all nonunique alignments, and enforces post-v1.3 Solexa quality scoring as well as using ten threads to speed execution:

```
$ bowtie -S -n 3 -m 1 --solexa1.3-quals -p 10 -l
80 $HG19 /path/to/file.fastq output.sam
```

To compress and sort the rows in the resulting SAM file (`output.txt`), the following command should be issued at the command-line:

```
$ samtools view -bS -o output.bam output.sam
$ samtools sort output.bam output.sorted
```

### 3.5 Proportion of Uniquely Mapped Reads

The number of reads that map uniquely to the reference genome also gives an indication of the quality of the data. This is a good test to carry out when several samples are available to indicate whether any sequencing anomalies may have occurred. Excluding the `-m` flag can dramatically increase the proportion of mapped reads as highlighted below. However, the number of uniquely mapped reads does not change.

```
$ bowtie -S -m 1 --solexa-quals -p 10 \
$HG19 SRR030769.fastq SRR030769.2.sam
# reads processed: 7173011
# reads with at least one reported alignment:
3891555 (54.25%)
# reads that failed to align: 536307 (7.48%)
# reads with alignments suppressed due to -m:
2745149 (38.27%)
Reported 3891555 alignments to 1 output stream(s)
```

```
$ bowtie -S --solexa-quals -p 10 \
$HG19 SRR030769.fastq SRR030769.4.sam
# reads processed: 7173011
# reads with at least one reported alignment:
6636704 (92.52%)
# reads that failed to align: 536307 (7.48%)
Reported 6636704 alignments to 1 output stream(s)
```

### 3.6 Quality Control

Sequencing performance varies and this can lead to strong biases for certain nucleotides. Moreover, read qualities are not constant across reads, gradually diminishing towards the 3′ end. To graphically capture this, the Python HTSeq library [24] has two scripts, `htseq-qa` and `htseq-count`, that may be invoked at the command-line. `htseq-qa` takes a FASTQ or SAM file (the output from mapping) and estimates the expected proportion of each nucleotide at each read position; it also plots the mean quality score at each read position (*see* Figs. 1 and 2). In the quality score plot, colour darkness is proportional to quality.

Figure 1 shows what a good short read dataset looks like, with all nucleotides uniformly represented along most of the read length. In contrast, Fig. 2 depicts variations and nucleotide biases produced using poor quality reads. Read qualities in the first figure are far higher than in the second. The command `htseq-qa` is invoked as follows:

```
$ htseq-qa -type=solexa-fastq <FASTQ_file>
$ htseq-qa <SAM_file>
```

and gives the results in a PDF file.

The `htseq-count` command takes a SAM and GFF annotation file to give the number of reads that map to each genomic feature. More information is available at the HTSeq website [24].

### 3.7 Identification of Heterozygous SNPs or Novel Variants

Once aligned, most sequence positions have a strong representation for one nucleotide. While not necessarily true, it is reasonable to consider that the weakly represented nucleotide is a possible sequencing error. In this way, a large number of SNP positions will be considered to be homozygous and are thus uninformative for

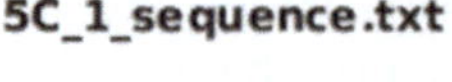


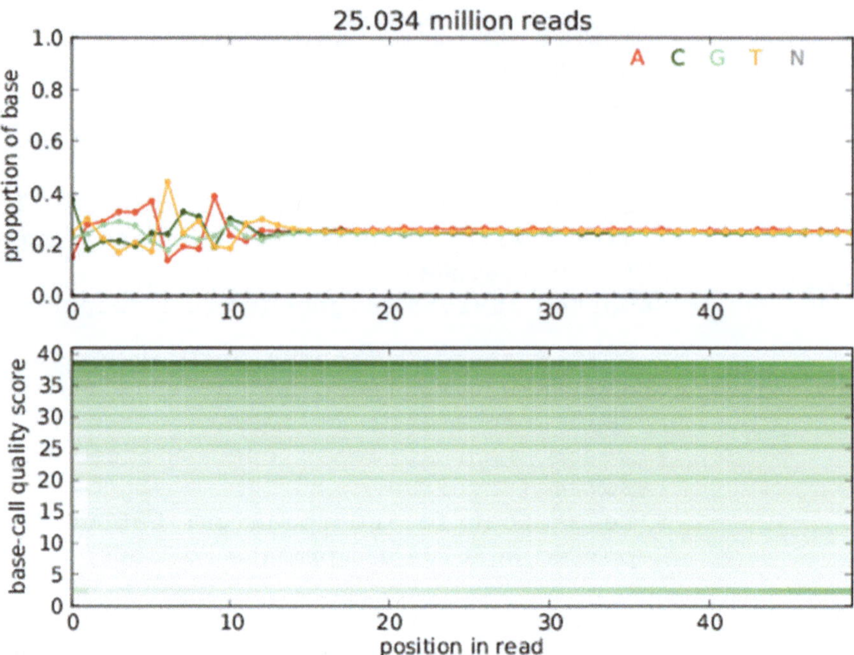


**Fig. 1** Example of high-quality short-read data. Graph generated using htseq-qa script

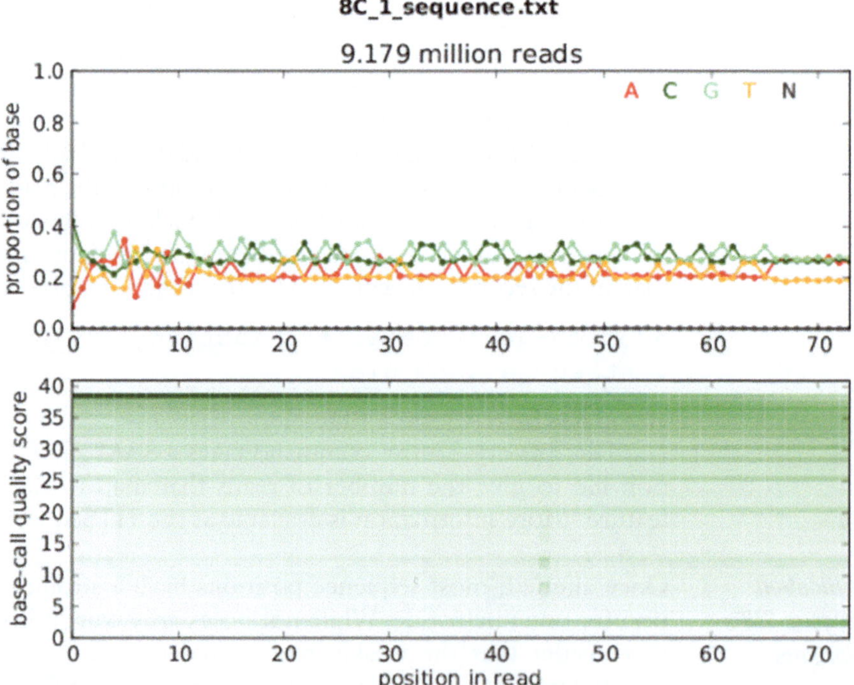


**Fig. 2** Example of a poor-quality short-read data. Graph generated using htseq-qa script

ASE inference. However, there might also exist novel variants that have not been classified as SNPs that show evidence of heterozygosity. The choice of whether to restrict identification only to known SNPs or to include novel variants affects the number of positions considered for ASE inference. On the other hand, if SNP data has not been characterized for the organism, then using novel variants will be the only option.

### 3.8 Pileup

ASE inference can be carried out using the `mpileup` function available with `SAMtools` [23]. The `mpileup` function requires a sorted BAM file (as was created in the previous step) and a reference genome in FASTA format.

```
$ samtools mpileup -C50 -f /path/to/FASTA/
file/ref.fa output.sorted.bam > columns.txt
```

The -C50 option tells the SAMtools pileup engine to modify quality scores for reads with excessive mismatches, whereas the -f option indicates that the next command line parameter is a reference FASTA file. The output file `columns.txt` contains all genomic positions in the alignment in pileup format. The fields in the pileup format are: chromosome name, coordinate, reference nucleotide, number of reads that overlap this position, encoded nucleotide information for each read in the alignment, and the corresponding quality scores. A few rows are displayed as below:

```
chr6 410512  T  25 .,,,,,,,.,.,.,,,,..,......^S.
""!#&%%%%%"&%!"!$%#%%!!"
chr6 410513 A 27 Gg,gg,,.,G,.,g,GG,GGG....^S.^SG
%$!!%#%%&!%!&!"$%%!%!%
chr6 410514  T  27 .$,,,,,,,.,.,.,,,,..,.........
"&!"&$%&&!#&%%!"!%!%%%&!#!$
chr6 410515  T  27 ,,,,,,,.,.,.,,,,..,.........^S.
%!#&%!%%%%!%!%!$&%&!"!$!
chr6 410516  G  27 ,$,,,,,,.,.,.,,,,..,.........
%!#&%$&%$!&%%!%$$$&&%!%!
```

Each symbol in the fifth column (and, correspondingly, in the last column) represents the nucleotide from one read (hence the reads have been "piled up" at the position represented by the second column). The fifth column is encoded relative to the reference nucleotide (third column). See Table 2 for details on how to interpret this encoding.

### 3.9 Filtering

The data that is generated at the pileup step consists of all aligned positions, the large majority of which are homozygous and thus uninformative for ASE. Filtering aims to eliminate, as objectively as possible, uninformative positions. The following are some of the methods that have been employed in several studies [25, 26]. We include short snippets of Python code that may be modified as needed.

**Table 2**
**Symbols used to encode a pileup**

| Symbol | Meaning |
|---|---|
| '.' | Matches to forward (+) strand |
| ',' | Matches to reverse (−) strand |
| '^' | Coincides with beginning of a read. The ASCII value of the character after the '^' minus 33 gives the mapping quality |
| '$' | Coincides with the end of a read |
| 'ACGTN' | A mismatch in on the forward strand |
| 'actgn' | A mismatch on the reverse strand. (*NB* this does not refer to the complement to the forward strand) |
| '>' or '<' | A reference skip |
| '\+[0-9]+[ACGTNacgtn]+' | An inserted nucleotide(s) between this reference position and the next reference position of length given by the preceding integer, e.g., '+2AC' means that there is an insertion consisting of 'AC' at the coordinate of that row for the respective read |
| '-[0-9]+[ACGTNacgtn]+' | Similar to above but applies for a deletion. The character '*' replaces the deleted nucleotides in aligned reads |

The pileup data (see above) consists of information on the chromosome, coordinate, reference allele, length and encoded nucleotide information, all of which is needed to perform most of these filtering steps. The fifth column needs to be decoded to determine the extent to which matching and mismatching occur. The following Python function can be used to do this:

```
def col5(col):
"""
given the string of the fifth column
returns match, mismatch, ambiguous
"""
plus = col.count(".") # count matches to + strand
minus = col.count(",") # count matches to
- strand
mismatch = {"A":0,"C":0,"G":0,"T":0}
mismatch["A"] = col.count("A") + col.count("a")
mismatch["C"] = col.count("C") + col.count("c")
mismatch["G"] = col.count("G") + col.count("g")
mismatch["T"] = col.count("T") + col.count("t")
     ambiguous = col.count("N") + col.count("n")
     return plus+minus, mismatch, ambiguous
```

The output of this function may then be processed to discriminate between positions based on several criteria as highlighted below. For example, the loop below prints out all positions with more than 20 reads and displays the chromosome, position, reference allele, number of each nucleotide, and number of ambiguous reads that align to that position.

```
# open the file
f = open("columns.txt","r")
# read each line one at a time
for l in f:
l2 = l.strip().split("\t") # convert each line
to a list of fields
match,mismatch,ambiguous = col5(l2[4])
total   =   sum(mismatch.values())   +   match   +
ambiguous
mismatch[l2[2]] += match
if total >= 20:
     print"\t".join([l2[0],l2[1],l2[2],l2[3]])
+ \
"\t".join(map(str,[mismatch["A"],mismatch["C"]]
,mismatch["G"], \
mismatch["T"],ambiguous]))
f.close()
```

The output of this is as shown below. The order of columns is given in the previous paragraph. Observe that most positions are homozygous as indicated by the strong representation for one nucleotide.

```
chrX    77381551        T       79      17      1
5       56      0
chrX    79965765        A       31      0       0
31      0       0
chrX    80533828        A       20      0       0
20      0       0
chrX    85339979        G       20      20      0
0       0       0
chrX    87352142        A       25      0       0
25      0       0
chrX    88735847        T       22      0       22
0       0       0
chrX    88735858        A       99      0       0
97      2       0
chrX    88735862        A       210     0       0
210     0       0
chrX    91368233        A       20      16      3
1       0       0
```

### *3.10 Read Depth and Second Most Abundant Allele Thresholding*

As seen above and will be demonstrated shortly, the power to infer ASE is, to a large extent, dependent on the number of reads [25, 27]. We, therefore, need to exclude all positions that fall below a certain read depth threshold or that demonstrate extreme imbalance, as these are better attributed to sequencing error. To do this, the read depth threshold may be set to, for example, 20 [7] or 50 [25], depending on the statistical power required (*see* Subheading 3.12). Additionally, a maximum threshold should be set for the detectable allelic imbalance by imposing a lower bound on the proportion of reads mapping to the lower expressed allele in the sample. For example, this lower bound was set to 15% by Heap et al. [25], corresponding to a maximum allelic bias of 5.67-fold and will be used for illustrative purposes here. This also implies that the second most abundant allele should not exceed 85% since a nonreference allele might be the most abundant. Below is a short Python script that implements this.

```
for l in f:
l2 = l.strip().split("\t")
match,mismatch,ambiguous = col5(l2[4])
total = sum(mismatch.values()) + match +
ambiguous
# get the second most abundant allele
ordered = mismatch.values()
ordered.sort()
# sort the mismatch frequencies
if ordered[3] >= .15*total and ordered[3] <=
.85*total and total >= 20:
     print"\t".join([l2[0],l2[1],l2[2],l2[3]])
+ \
"\t".join(map(str,[mismatch["A"],mismatch["C"]
,mismatch["G"], \
mismatch["T"],ambiguous]))
```

After applying these filters, the remaining lines of data are shown below. All positions are non-homozygous (heterozygous or false positives).

```
chr22   39713463   T   31    0   9
1       21         0
chr22   39714556   G   145   1   1
95      48         0
chr22   39916631   G   27    0   8
19      0          0
chr22   39916632   C   27    0   19
8       0          0
chr22   39916635   T   24    0   4
0       20         0
chr22   39916636   C   25    0   21
0       4          0
```

```
chr22   39916637        A       24      20      4
0       0       0
chr22   41469837        C       25      0       18
0       7       0
```

While the above constraints are sufficient to eliminate homozygous positions, there still remain potential false positives. For example, the genomic region encoding the major histocompatibility genes in vertebrates contains a large number of polymorphic positions and some of these may be heterozygous. Since they occur quite close to each other, they may cause mapping bias and should thus be excluded.

Three other constraints suggested before final analysis are [25]:

1. To test whether the allelic call is dependent on the position along the read, one may use the Kolmogorov–Smirnov test between the distribution of allelic calls (reference and nonreference) and that of positions of SNPs/novel variants along the read's length. All polymorphic positions that are significantly dependent on position at the $\alpha = 0.01$ level should be excluded.
2. $\chi 2$ goodness-of-fit test to eliminate strand bias. Positions that have a bias to a particular strand significant at the $\alpha = 0.01$ level should be excluded.
3. Indels distort coordinates close to them. All SNPs/novel variants that are situated within one read length on either sided of the indel should also be excluded since they are likely to simulate polymorphisms.

Since the `mpileup` function is also available in various language bindings (C/C++, Python, Perl, Java, Ruby, and Lisp), users can take advantage of this to combine both the pileup step and filtering in one script. See the `SAMtools` website for details [23].

### 3.11 Using snpMatrix as a Pre-Filter (Optional)

A good number of novel variants are likely to be false positives as a result of, for example, genotyping errors. False positives need to be eliminated in order to increase the confidence of ASE inference. One way to do this is to filter using well-known SNPs and a possible source of such SNPs is dbSNP. However, it has been observed that the use of dbSNP SNPs as a filter for polymorphic sites leads to a higher false discovery rate in human studies compared to filtering using HapMap SNPs [25]. We detail the use of HapMap SNPs as a prefilter using the R package `snpMatrix`. In our example, we first download the Yoruba SNP data from the HapMap database and assemble SNPs from each chromosome into a single file.

```
library(snpMatrix)
chr   <-   c(as.character(1:22),'X','Y','M')   #
replace with chromosome numbers
for (i in chr) {
fn <- paste("YRI.chr",i,".hapmap",sep="")
```

```
cat(paste("Filename: ",fn),"\n") # keep the user
informed
url <- paste("http://hapmap.ncbi.nlm.nih.gov/
downloads/\
genotypes/latest_phaseII+\
III_ncbi_b36/forward/non-redundant/
genotypes_chr",i,\
"_YRI_r27_nr.b36_fwd.txt.gz",sep="")
cat(paste("URL: ",url),"\n") # keep the user
informed
result <- read.HapMap.data(url)
# put the data into a file
write.table(result$snp.support,file=fn, quote=F,
sep="\t")
cat(paste("Written SNP data for chromosome ",i,"
for YRI \ (Yoruba)",sep=""),"\n")
rm(result) # free the result variable
}
```

Next, for each chromosome, we filter all non-SNPs from the pileup data we obtained above. Here, we assume that pileup data is separated into one file per chromosome (e.g., `chr1.snp`).

```
chr <- c(as.character(1:22),'X','Y','M') #
replace with chromosome numbers
snps <- data.frame()
get.SNPs <- function(i,path.snp,path.YRI) {
# positions with reads > 20 per chromosome
fn <- paste(path.snp,"chr",i,".snp",sep="")
snp.rough <- data.frame(read.delim(fn))
names(snp.rough) <- c("chr","pos","ref","count"
,"A","C","G","T","N")
# HapMap SNPs for YRI
fn2 <- paste(path.YRI,"YRI.chr",i,".
hapmap",sep="")
snp.chr <- data.frame(read.delim(fn2,header =
T))
# now, merge and only retain those in both
snps.proper <- merge(snp.chr, snp.rough,by.
x="Position",by.y="pos")
snps.proper
}
path.snp <- "./<RNA-seq sample i>/" # replace
with the appropriate folder name
path.YRI <- "./SNPs/"
for (i in chr){
snps <- rbind(snps,get.SNPs(i,path.snp,path.YRI))
}
```

We can now construct a data frame that will hold the prefiltered SNPs.

```
snps.AI <- snps[,1:8]
for (i in c(1:length(snps.AI$count))) {
x    <-    c(snps[i,7],    snps[i,8],    snps[i,9],
snps[i,10])
y        <-        x[sort.list(x,partial=NULL,na.
last=T,decreasing=T)]
snps.AI[i,7] <- y[1]
snps.AI[i,8] <- y[2]
 }
# rename the last two columns
names(snps.AI)[7:8] <- c('A1','A2')
```

These still have to be further filtered down to only those SNPs that meet the imbalance conditions. However, this method returns significantly fewer results because some heterozygous positions are not SNPs. At the R prompt, we may now filter out homozygous positions.

```
> snps.AI[snps.AI$A2 >= .15*snps.AI$count,]
      Position dbSNPalleles Assignment Chromosome
Strand count A1 A2
238 150496685          C/T         C/T        chr5
+    32 21  7
```

### 3.12 Power to Infer ASE

In the previous step, we discussed criteria to filter positions based on three conditions. Two of these constraints placed a bound on the expected extent of allelic bias. Furthermore, we separated the identified positions based on the number of reads that mapped to each position. Intuitively, the greater the number of mapped reads, the higher the power to detect ASE. Let $A$ be the rate of transcription from the more highly expressed allele $A_1$ divided by that of the lower expressed allele $A_2$. We assume that only two alleles are present and the other nucleotides have very low counts, corresponding to sequencing errors. The probability of observing $N = n_{A_1}$ of $C$ reads mapping to $A_1$ is given by the binomial distribution as:

$$P\left(N = n_{A_1}\right) = \binom{C}{n_{A_1}}\left(\frac{A}{A+1}\right)^{n_{A_1}}\left(\frac{1}{A+1}\right)^{(C-n_{A_1})} \tag{1}$$

For large $C$ this may be approximated by a normal distribution. For a particular value of $A$ using $\alpha$ of 5%, the power to detect ASE may be plotted against $C$, the number of reads as shown in Fig. 3. As expected, detection of subtle ASE (1.25-2) requires more reads, whereas stronger ASE (5–100) can be detected with far fewer reads [27].

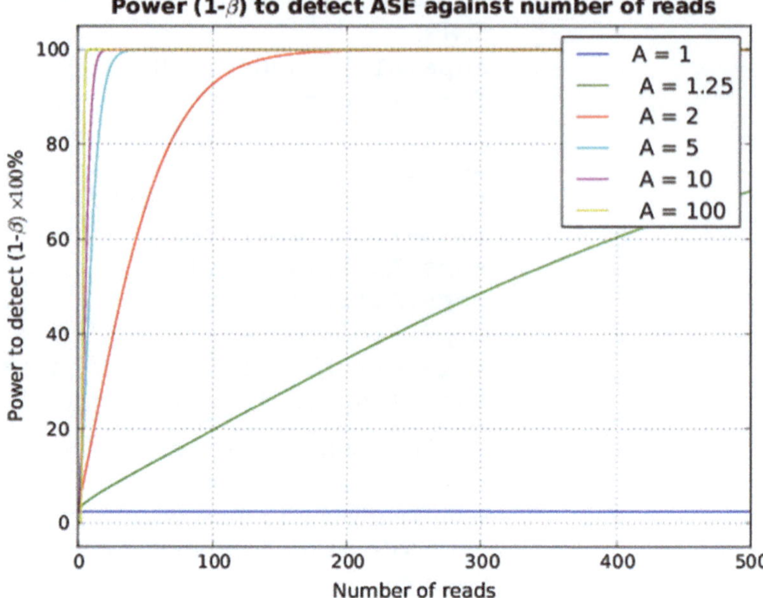


**Fig. 3** Graph of variation of statistical power with read depth. Graph generated using htseq-qa script

### *3.13 Estimation of Gene Expression from RNA-seq Data*

RNA-seq data is most frequently used to quantify expression levels of transcripts (e.g., [28–30]). We take a brief detour to illustrate how this is done though it is not a necessary step in ASE inference but may be carried out alongside it. One of the most straightforward methods involves counting the number of reads that align to exons. These counts are normalized by the combined length of the exons of a gene (in kilobases) and by the total number of mapped reads (in millions) to give reads per kilobase per million mapped reads (RPKM). If $r_i$ reads map to the exons of gene $i$ then:

$$RPKM_i = \frac{1}{mg} r_i$$

is an expression metric, where $m$ is the total number of mapped reads in millions and $g$ is the sum of all exon lengths (in kilobases) [28].

The two-sided binomial test is the simplest test for allelic imbalance [7].

```
snps.binom <- list()
length(snps.binom) <- length(snps.AI$A1)
for (i in c(1:length(snps.AI$Position))) {
snps.binom[i]   <-   binom.test(c(snps.AI[i,7],
snps.AI[i,8]),\
p=0.5)$p.value
 }
```

The main weakness of this test, however, is that it does not account for sequencing errors that inevitably appear. An alternative method is to infer ASE using a model that directly incorporates sequencing errors (allele miscalls). This method uses a likelihood ratio test to infer ASE by evaluating a likelihood function at each position using two parameters that are empirically determined from the RNA-seq data and the SNP information: the probability, $\pi$, that a call from a given read is wrong and a conditional probability $\pi_{X,Y} = P(\text{Called} = X | \text{True} = Y)$ of calling a nucleotide $X$ when in fact it should be a $Y$ [31]. For a heterozygous position with alleles $A_1$ and $A_2$, we denote $M_1$ and $M_2$ as the alternative (incorrect) nucleotides, and $c_X$ as the number of reads that correspond to each nucleotide $X$ (e.g., for the number of $A_1$ reads). We also denote $f_{A_1}$ and $f_{A_2}$ as the actual frequencies of $A_1$ and $A_2$ among all transcripts such that $f_{A_1} + f_{A_2} = 1$. We can now write the log-likelihood of $f = f_{A_1}$ as:

$$
\begin{aligned}
\ln L(f) = K &+ c_{A_1} \cdot \ln\left(f(1-\pi) + (1-f) \cdot \pi \cdot \pi_{A_1,A_2}\right) \\
&+ c_{A_2} \cdot \ln\left((1-f)(1-\pi) + f \cdot \pi \cdot \pi_{A_2,A_1}\right) \\
&+ c_{M_1} \cdot \ln\left(\pi\left[f \cdot \pi_{M_1,A_1} + (1-f) \cdot \pi_{M_1,A_2}\right]\right) \\
&+ c_{M_2} \cdot \ln\left(\pi\left[f \cdot \pi_{M_2,A_1} + (1-f) \cdot \pi_{M_2,A_2}\right]\right)
\end{aligned}
$$

Setting $\pi = 0$ reduces this to the binomial case. The likelihood ratio test for allelic imbalance will then be:

$$
G\left(c_{A_1}, c_{A_2}, c_{M_1}, c_{M_2}\right) = -2 \ln \frac{L\left(\frac{1}{2}\right)}{\max_f L(f)}
$$

which, under the null hypothesis of equal allelic expression, follows a $\chi 2$ distribution with one degree of freedom. The MLE of $L(f)$ may be obtained by numerical optimization in R. Estimation of values for $\pi$ and $\pi_{X,Y}$ requires that data for a sufficient number of SNPs be available from the RNA-seq reads [31]. To estimate $\pi_{X,Y}$, we use miscalls $X$ at all homozygous sites $YY$ since the bias for $Y$ indicates that the $X$'s are errors. On the other hand, $\pi$ may be estimated by getting the average maximum likelihood estimate for (1) at each SNP.

### 3.14 Correcting for Multiple Testing

Given that separate hypothesis tests are carried out for each potential ASE position, it is necessary to implement methods to correct for multiple hypothesis testing. This is most often achieved through control of the false discovery rate (FDR). Several methods to control the FDR are available in the R statistical computing environment [32].

### 3.15 Annotation

Finally, it is often useful to obtain annotation data for loci that show evidence of ASE following correction for multiple testing. There are several R packages that may be used to do this and it is left to the reader to consult their respective manuals.

1. `snpMatrix` is used for annotating SNPs. It is only useful for organisms for which SNP genotype data are available and filtering using the obtained SNPs will eliminate unannotated polymorphic sites.
2. `GenomicRanges` makes it easy to handle genomic coordinates and can be applied to isolating SNPs by loci.
3. `GenomicFeatures` for annotation by genomic features (exons, transcripts, or genes).
4. `org.Xx.yy.db` for gene-centric annotation, where `Xx` is the Genus and species of the organism and `yy` is the source name of the data's central ID [33].

### 3.16 Validation

There are two principal methods by which ASE inference results may be validated experimentally. The first involves allele-specific reverse-transcription qPCR (AS RT-qPCR). AS RT-qPCR can be performed on an Applied Biosystems 7900HT Sequence Detection System as described in [26] or in Chap. 4 of this volume by Day and Macknight, as well as other platforms. Users should refer to the manufacturer's protocol that is available for the equipment. Alternatively, since ASE signals indicate the presence of imprinting or *cis*-eQTLs, where these have been mapped using alternative methods (e.g., microarrays), results may then be compared to verify ASE signals detected using RNA-seq [34].

## 4 Comparison with Other Methods for ASE Inference

The analysis of RNA-seq as presented here is not the only method by which ASE may be inferred. However, it presents a method that provides flexibility to carry out other analyses alongside it. There are alternative methods of inference, some of which are specialized, that we briefly describe here.

1. SNP-arrays are high-density microarrays whose probes consist of oligonucleotides containing well-characterized polymorphic positions [35, 36]. Their use, however, is not restricted to ASE and they are routinely used for high-throughput genotyping. SNP-arrays have been applied to global ASE inference studies in both humans [1] and plants [10].
2. Use of padlock probes is an alternative polymorphism-targeted method in which known exonic polymorphisms can be captured on a large scale [37, 38]. Padlock probes are highly specific and scalable and their application to ASE is described in [6].

3. In massively parallel signature sequencing (MPSS), cDNA libraries are amplified then polymerized onto beads containing uniquely labeled segments [39]. MPSS has been applied to infer ASE in maize meristem [8].

## 5 Conclusions

Expression imbalance between alternative alleles in diploid organisms or organisms of higher ploidy has the capacity to reveal important information about the control of gene expression and can, in particular, be used to identify *cis*-regulatory variants. While RNA-seq is still a relatively new and developing technology, it has the capacity to play a central role in transcriptomics. The analysis of RNA-seq data is challenging and caveats associated with this technology depend on the specific application. In the case of the inference of ASE, even a relatively small proportion of reads that are incorrectly mapped to the reference genome sequence can give strong false-positive signals and very misleading results. The sequence of analysis steps outlined here provides just one possible analysis pipeline and set of analytical choices that could be used to infer ASE from RNA-seq data and many variants are possible. As the technologies mature, a greater consensus around the optimal analysis strategies is also likely to emerge.

### References

1. Ge B, Pokholok DK, Kwan T, Grundberg E, Morcos L, Verlaan DJ, Le J, Koka V, Lam KCL, Gagné V et al (2009) Global patterns of *cis* variation in human cells revealed by high-density allelic expression analysis. Nat Genet 41:1216–1222
2. Pastinen T (2010) Genome-wide allele-specific analysis: insights into regulatory variation. Nat Rev Genet 11:533–538
3. Wagner JR, Ge B, Pokholok D, Gunderson KL, Pastinen T, Blanchette M (2010) Computational analysis of whole-genome differential allelic expression data in human. PLoS Comp Biol 6:e1000849
4. Serre D, Gurd S, Ge B, Sladek R, Sinnett D, Harmsen E, Bibikova M, Chudin E, Barker DL, Dickinson T et al (2008) Differential allelic expression in the human genome: a robust approach to identify genetic and epigenetic *cis*-acting mechanisms regulating gene expression. PLoS Genet 4:e1000006
5. Majewski J, Pastinen T (2011) The study of eQTL variations by RNA-seq: from SNPs to phenotypes. Trends Genet 27:72–79
6. Zhang K, Li JB, Gao Y, Egli D, Xie B, Deng J, Li Z, Lee JH, Aach J, Leproust EM et al (2009) Digital RNA allelotyping reveals tissue-specific and allele-specific gene expression in human. Nat Methods 6:613–618
7. Degner JF, Marioni JC, Pai AA, Pickrell JK, Nkadori E, Gilad Y, Pritchard JK (2009) Effect of read-mapping biases on detecting allele-specific expression from RNA-sequencing data. Bioinformatics 25:3207–3212
8. Guo M, Yang S, Rupe M, Hu B, Bickel DR, Arthur L, Smith OG-w (2008) Genome-wide allele-specific expression analysis using massively parallel signature sequencing (MPSS™) reveals *cis*- and *trans*-effects on gene expression in maize hybrid meristem tissue. Plant Mol Biol 66:551–563
9. Cookson W, Liang L, Abecasis G, Moffatt M, Lathrop M (2009) Mapping complex disease traits with global gene expression. Nat Rev Genet 10:184–194
10. Zhang X, Borevitz JO (2009) Global analysis of allele-specific expression in *Arabidopsis thaliana*. Genetics 182:943–954
11. Jiménez-Gómez JM, Wallace AD, Maloof JN (2010) Network analysis identifies ELF3 as a QTL for the shade avoidance response in *Arabidopsis*. PLoS Genet 6:e1001100

12. West MAL, Kim K, Kliebenstein DJ, Van Leeuwen H, Michelmore RW, Doerge RW, Clair DAS (2007) Global eQTL mapping reveals the complex genetic architecture of transcript-level variation in *Arabidopsis*. Genetics 175:1441–1450
13. Wittkopp PJ, Haerum BK, Clark AG (2004) Evolutionary changes in *cis* and *trans* gene regulation. Nature 430:85–88
14. Keurentjes JJB, Fu J, Terpstra IR, Garcia JM, Van Den Ackerveken G, Snoek LB, Peeters AJM, Vreugdenhil D, Koornneef M, Jansen RC (2007) Regulatory network construction in *Arabidopsis* by using genome-wide gene expression quantitative trait loci. Proc Natl Acad Sci U S A 104:1708–1713
15. Metzker ML (2009) Sequencing technologies–the next generation. Nat Rev Genet 11:31–46
16. Wang Z, Gerstein M, Snyder M (2009) RNA-Seq: a revolutionary tool for transcriptomics. Nat Rev Genet 10:57–63
17. Ozsolak F, Milos PM (2010) RNA sequencing: advances, challenges and opportunities. Nat Rev Genet 12:87–98
18. Trapnell C, Salzberg SL (2009) How to map billions of short reads onto genomes. Nat Biotech 27:455–457
19. Gilad Y, Pritchard JK, Thornton K (2009) Characterizing natural variation using next-generation sequencing technologies. Trends Genet 25:463–471
20. Gibbs RA, Belmont JW, Hardenbol P, Willis TD, Yu F, Yang H, Ch'ang LY, Huang W, Liu B, Shen Y et al (2003) The international HapMap project. Nature 426:789–796
21. Leinonen R, Sugawara H, Shumway M (2011) The sequence read archive. Nucleic Acids Res 39:D19–D21
22. Langmead B, Trapnell C, Pop M, Salzberg SL et al (2009) Ultrafast and memory-efficient alignment of short DNA sequences to the human genome. Genome Biol 10:R25
23. Li H, Handsaker B, Wysoker A, Fennell T, Ruan J, Homer N, Marth G, Abecasis G, Durbin R et al (2009) The sequence alignment/map format and SAMtools. Bioinformatics 25:2078–2079
24. Anders, S. HTSeq: Analysing high-throughput sequencing data with Python. http://www-huber.embl.de/users/anders/HTSeq/doc/overview.html. Accessed 30 Jan 2013.
25. Heap GA, Yang JHM, Downes K, Healy BC, Hunt KA, Bockett N, Franke L, Dubois PC, Mein CA, Dobson RJ et al (2010) Genome-wide analysis of allelic expression imbalance in human primary cells by high-throughput transcriptome resequencing. Hum Mol Genet 19:122–134
26. Tuch BB, Laborde RR, Xu X, Gu J, Chung CB, Monighetti CK, Stanley SJ, Olsen KD, Kasperbauer JL, Moore EJ et al (2010) Tumor transcriptome sequencing reveals allelic expression imbalances associated with copy number alterations. PLoS One 5:e9317
27. Fontanillas P, Landry CR, Wittkopp PJ, Russ C, Gruber JD, Nusbaum C, Hartl DL (2010) Key considerations for measuring allelic expression on a genomic scale using high-throughput sequencing. Mol Ecol 19:212–227
28. Mortazavi A, Williams BA, McCue K, Schaeffer L, Wold B (2008) Mapping and quantifying mammalian transcriptomes by RNA-Seq. Nat Methods 5:621–628
29. Pickrell JK, Marioni JC, Pai AA, Degner JF, Engelhardt BE, Nkadori E, Veyrieras JB, Stephens M, Gilad Y, Pritchard JK (2010) Understanding mechanisms underlying human gene expression variation with RNA sequencing. Nature 464:768–772
30. Montgomery SB, Sammeth M, Gutierrez-Arcelus M, Lach RP, Ingle C, Nisbett J, Guigo R, Dermitzakis ET (2010) Transcriptome genetics using second generation sequencing in a Caucasian population. Nature 464: 773–777
31. Nothnagel M, Wolf A, Herrmann A, Szafranski K, Vater I, Brosch M, Huse K, Siebert R, Platzer M, Hampe J et al (2011) Statistical inference of allelic imbalance from transcriptome data. Hum Mutat 32:98–106
32. Team R. (2010) R: A language and environment for statistical computing. R Foundation for Statistical Computing Vienna Austria, (01/19).
33. Gentleman RC, Carey VJ, Bates DM, Bolstad B, Dettling M, Dudoit S, Ellis B, Gautier L, Ge Y, Gentry J et al (2004) Bioconductor: open software development for computational biology and bioinformatics. Genome Biol 5:R80
34. Babak T, Garrett-Engele P, Armour CD, Raymond CK, Keller MP, Chen R, Rohl CA, Johnson JM, Attie AD, Fraser HB et al (2010) Genetic validation of whole-transcriptome sequencing for mapping expression affected by *cis*-regulatory variation. BMC Genomics 11:e473
35. Fan JB, Oliphant A, Shen R, Kermani BG, Garcia F, Gunderson KL, Hansen M, Steemers F, Butler SL, Deloukas P et al (2003) Highly parallel SNP genotyping. Cold Spring Harbor Symp Quant Biol 68:69–78
36. Fan JB, Chee MS, Gunderson KL (2006) Highly parallel genomic assays. Nat Rev Genet 7:632–644
37. Hardenbol P, Banér J, Jain M, Nilsson M, Namsaraev EA, Karline-Neumann GA, Fakhrai-Rad H, Ronaghi M, Willis TD, Landegren U, Davis RW (2003) Multiplexed genotyping with sequence-tagged molecular inversion probes. Nat Biotech 21:673–678

38. Hardenbol P, Yu F, Belmont J, MacKenzie J, Bruckner C, Brundage T, Boudreau A, Chow S, Eberle J, Erbilgin A et al (2005) Highly multiplexed molecular inversion probe genotyping: over 10,000 targeted SNPs genotyped in a single tube assay. Genome Res 15:269–275
39. Brenner S, Johnson M, Bridgham J, Golda G, Lloyd DH, Johnson D, Luo S, McCurdy S, Foy M, Ewan M et al (2000) Gene expression analysis by massively parallel signature sequencing (MPSS) on microbead arrays. Nat Biotech 18:630–634

# Chapter 5

# Screening for Imprinted Genes Using High-Resolution Melting Analysis of PCR Amplicons

Robert Day and Richard Macknight

## Abstract

High-resolution melting (HRM) analysis is a technique that enables researchers to detect polymorphisms in DNA molecules based on different melting profiles and is becoming widely used as a method for detecting SNPs in genomic DNA. In this chapter, we describe how HRM analysis can be used to detect allelic imbalances typical of imprinted genes, where alleles are differentially expressed based on their parent of origin. This involves first producing hybrid seed using parental plants that have sufficient genetic differences to distinguish the parental origin of each allele of the candidate genes. RNA is then isolated from the hybrid seed and converted to cDNA. PCR amplicons are produced using primers designed to span a polymorphic sequence within the transcript of the candidate gene. By using a real-time PCR machine with HRM analysis capability, the PCR amplicons can be analyzed without further manipulations directly after amplification to detect instances of strong allelic imbalance and parent-of-origin-dependent expression.

**Key words** High-resolution melting, Amplicon analysis, Allelic imbalance, Imprinting, Parent of origin

## 1 Introduction

Imprinting is an epigenetic phenomenon whereby one allele of a gene is exclusively or predominantly expressed depending on its parental origin. In flowering plants imprinting primarily occurs in the endosperm, a triploid fertilization product that acts to support the developing embryo. Imprinted genes are predicted to encode key regulators of endosperm development and seed size [1, 2]. Until recently, only a few imprinted genes were known in plants. However, the use of novel approaches has lead to the identification of many more [3–8]. The most successful approach has been to use short-read deep sequencing of transcript libraries generated from immature seed or endosperm to identify imprinted genes [4, 5, 8, 9]. However, this approach is inefficient at assaying low expressed genes. Here, we describe the use of high-resolution melting (HRM) analysis to identify imprinted genes (see also the following chapter of McKeown et al. concerning quantification of such

Charles Spillane and Peter C. McKeown (eds.), *Plant Epigenetics and Epigenomics: Methods and Protocols*,
Methods in Molecular Biology, vol. 1112, DOI 10.1007/978-1-62703-773-0_5, © Springer Science+Business Media New York 2014

imprinted genes, Chap. 6). This method provides a rapid, cheap and reliable means of both screening for novel imprinted genes and investigating the conservation of imprinting across different accessions or species.

The first step involves choosing two accessions or varieties that are sufficiently different so that polymorphisms within the candidate genes are easily identified. The arabidopsis project that is well on its way to sequencing the genome of 1001 different accessions [10], and similar efforts in other species, has simplified this task. We have recently identified imprinted genes from hybrid crosses between the two arabidopsis accessions Columbia (Col-0) and Landsberg *erecta* (L*er*-0) and these crosses are shown in the summary diagram Fig. 1. In addition to producing reciprocal crosses between the two accessions to produce hybrid seed, it is also important to obtain seed from self-crossed plants. RNA can then be isolated from either pure endosperm or whole seed tissue and cDNA synthesized [11, 12]. To obviate the need for isolating endosperm away from the other seed tissues, we targeted genes that are predominantly expressed in the endosperm. These first steps are shared between many approaches to identify imprinted genes.

In this protocol, we use HRM analysis to determine if a candidate gene is biallelically expressed or if there is a strong parental bias in allelic expression. HRM analysis is an efficient method for genotyping DNA samples based on the altered melting characteristics of polymorphic PCR amplicons. During PCR, complimentary DNA strands randomly reanneal after each denaturing cycle. In an amplicon population that only contains products generated from one allele of the targeted gene only homodimers are formed. However, if two alleles are present, their polymorphic strands anneal to form heterodimers. Real-time PCR machines from various companies have the capability to monitor the proportion of DNA that denatures (or melts) at a range of temperatures. Heterodimeric populations generally have melting characteristics that deviate significantly from the homodimeric populations and are therefore easily distinguished. HRM analysis relies on the ability to monitor the melting of DNA as it happens. This is done by using intercalating dyes that are specific for double-stranded (ds) DNA and only fluoresce when bound. Therefore, in the presence of dsDNA, the fluorescent signal is detected, but as the temperature is increased and the dsDNA begins to melt, the signal decreases. When only single-stranded DNA remains, no signal is detected. Typical melting curves and how these are manipulated by the analysis software to reveal differences are shown in Fig. 2. We used the false hybrid melting trace, generated by mixing cDNA from the homozygous samples in a 1:1 ratio, as the baseline when viewing the melt difference plots. Figure 3 shows the HRM curves for three different genes. In all examples shown, the parental melting curves differ significantly from the false hybrid. For the biallelically

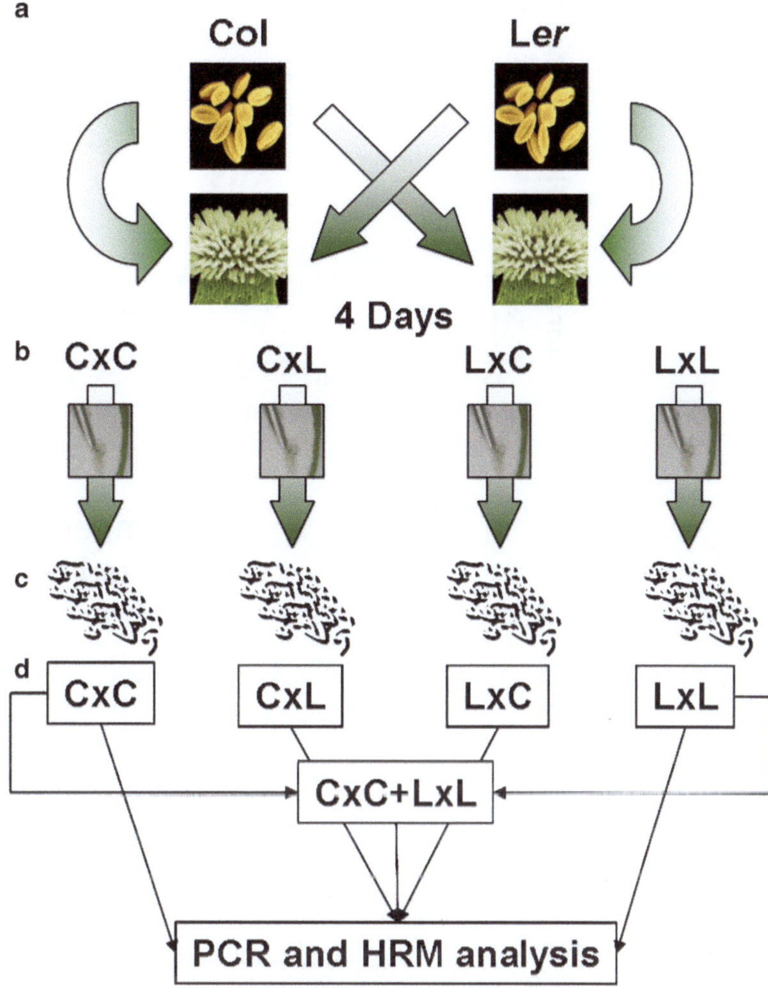


**Fig. 1** Summary of steps required to identify imprinted genes. (**a**) Two arabidopsis accessions, e.g., Col and L*er*, are reciprocally crossed and self-crossed to generate seeds with four different genotypes. (**b**) Immature seeds are dissected from the siliques 4 days after pollination. (**c**) RNA is isolated from the four samples and cDNA synthesized. A false hybrid is made by combining homozygous Col and homozygous L*er* cDNA in a 1:1 ratio (C × C + L × L). (**d**) Candidate genes are PCR amplified from the five cDNA samples and evaluated using HRM analysis. *LC* L*er* × Col, *CL* Col × L*er*, *C* × *C* Col self-cross, *L* × *L* L*er* self-cross

expressed gene, the melting curves from Col × L*er* reciprocal crosses are similar to the false hybrid (*see* Fig. 3a), whereas for the imprinted genes, the melting curves from Col × L*er* reciprocal crosses segregated with the homodimeric parental melting curves, indicating mono-allelic expression (*see* Fig. 3b). In the example of an imprinted gene shown in Fig. 3c, the melting curves for the parents differ and therefore which of the alleles is expressed can

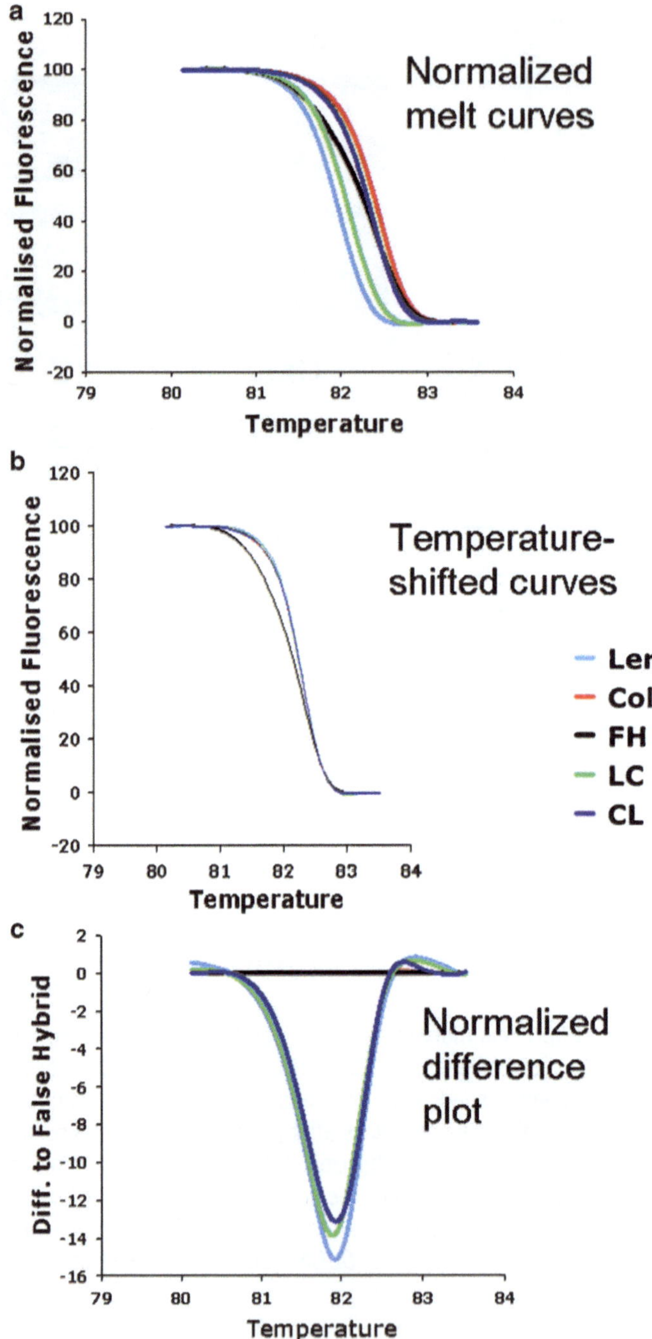


**Fig. 2** HRM data manipulations performed using the Roche LC480 Gene Scanning Analysis Software. (**a**) Normalization: the premelt fluorescence and postmelt fluorescence signals are set to 100 % and 0 %, respectively. (**b**) Temperature shifting: to easily distinguish homozygous and heterozygous SNPs, the normalized curves are shifted along the temperature axis to unify the point where all the dsDNAs are denatured. (**c**) Difference plot: subtracting the curves from a reference curve (for our analyses the False Hybrid sample is used) reveals differences in the shape of the melting curves. Different classes of samples are then apparent. *FH* false hybrid, *LC* L*er*× Col; CL, Col × L*er*

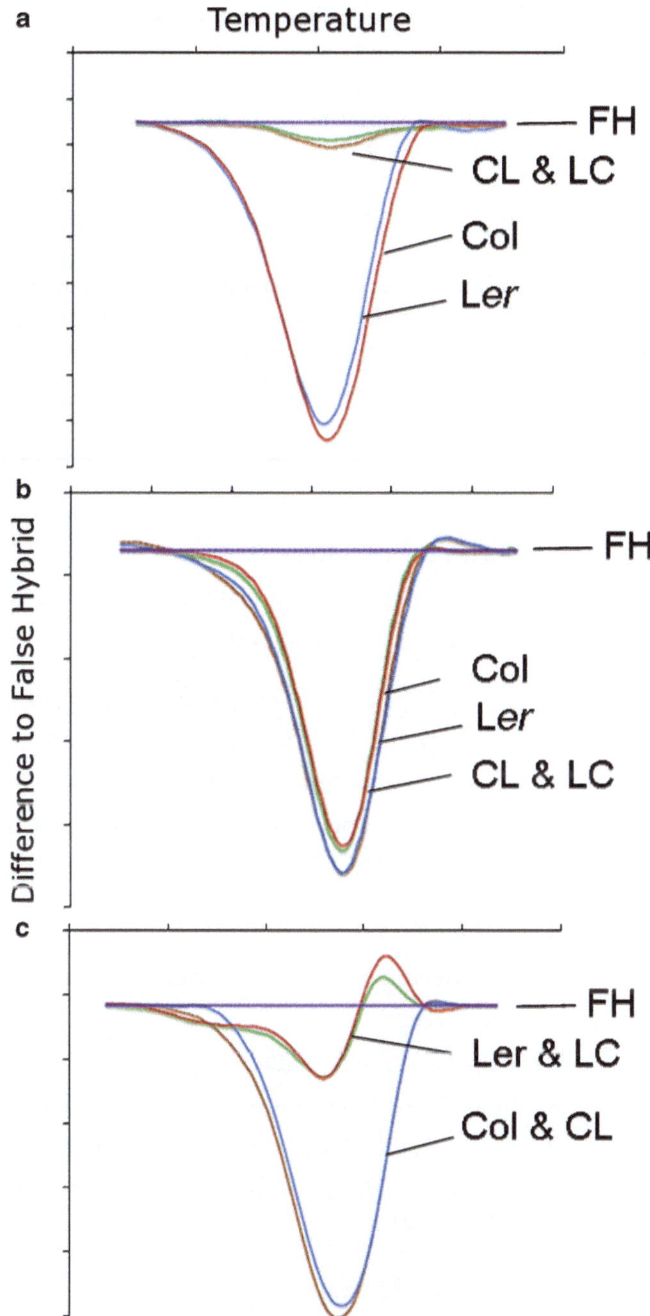


**Fig. 3** Examples of melting curves derived from three different genes. (**a**) Biallelically expressed gene where the melting curve of the CL and LC reciprocal crosses is similar to the false hybrid (FH) and not the parents. (**b**) An imprinted gene where the CL and LC reciprocal crosses are similar to the parents and not the false hybrid. (**c**) An imprinted gene where the melting curves of the parents differ, thereby revealing direction of imprinting (i.e., that only the maternally inherited allele is expressed). *FH* false hybrid, *LC* L*er* × Col; CL, Col × L*er* (*Note*: the mother is always written first)

also be determined. The differences between the melting curves of amplicons formed from RNA populations expressing only one allele of a gene compared to biallelic expression is usually detected by HRM without the need for assay optimization or the use of specialist primers or even reagents. However, differences in the melting profile can be accentuated by using saturating dyes and by using probes in combination with specialized primers (reviewed in ref. 13).

In this chapter, we provide a comprehensive guide to screening for allelic imbalance by HRM that will enable researchers to discover and/or validate parent-of-origin-dependent expression of candidate imprinted genes.

## 2 Materials

### 2.1 Primers

1. Nuclease-free water.
2. Nuclease-free 96-well plate or tubes for primer dilution.
3. Filter tips for micropipettes.

### 2.2 Template Preparation

1. RNaseZap (Sigma). CAUTION! RNaseZap is a mild irritant and so eye protection and gloves are recommended.
2. RNeasy Plant Mini Kit (Qiagen Cat. No. 74904).
3. Analytical grade absolute ethanol. CAUTION! Ethanol is flammable and a mild contact hazard so use with appropriate ventilation wear gloves and eye protection.
4. Sterile plastic Petri dish.
5. Hypodermic needle on syringe.
6. Fine forceps.
7. Scalpel with new blade.
8. Bioanalyzer RNA 6000 Nano Kit (Agilent 5067-1511).
9. RNase-free DNase I (Invitrogen Cat. No. 18068-015).
10. Superscript VILO cDNA Synthesis Kit (Invitrogen Cat. No. 11754-050).

### 2.3 Assay Setup by Robot and HRM-PCR

1. Nuclease-free water.
2. Nuclease-free 384-well plate or tubes for primer dilution.
3. Filter tips for micropipettes.
4. LightCycler 480 Multiwell Plate 384, white with sealing foils (Roche Cat. No. 04 729 749 001).
5. LightCycler 480 High-Resolution Melting Master (Roche Cat. No. 04 909 631 001).

## 3 Methods

A flow diagram of the steps required for a typical HRM screen is given in Fig. 1. Here, we cover the process from plants to data analysis. The PCR setup described here is designed for rapidly screening a large number of genes using conditions that we have found to give consistently good results for a wide range of amplifications. However, if an assay is being designed to target a particular gene, the user may want to optimize PCR conditions (*see* **Note 1**). Our screening approach uses 384-well plates, a liquid handling robot, and integration of HRM into the end of the PCR program. However, manual setup is suitable for the analysis of a low number of genes. It is also possible to carry out PCR and melt analysis separately (*see* **Note 2**).

### 3.1 Primers

We found Primer 3 primer design software to be amenable for HRM primer design [14]. As a first step, suitable SNPs need to be identified in the coding region of the targeted gene. While this aspect was once a major limiting factor to HRM screening, many millions of interaccession SNPs are now available thanks to initiatives such as the 1001 genomes project [10].

PCR primers should be designed based on the following criteria (1) the predicted PCR product should span a single SNP, (2) the PCR product should optimally be 150 bp in length but within the range of 50–350 bp, and (3) at least one of the primers should be designed to anneal over an intron–exon junction (*see* **Note 3**). Most genes will have more than one coding SNP available so if no suitable primer pairs can be designed try an alternative SNP position. If an alternative SNP is not available, the size of the amplicon can be increased, but it is unlikely that melt differences will be easily apparent in amplicons larger than 500 bp.

For HRM, it is recommended that HPLC-purified primers are used. When submitting primers for synthesis by your supplier, be aware that plates of 96 primers tend to be much less expensive than ordering individual tubes. Some suppliers will also apply the lower pricing for half plates (i.e., 48 oligos). A significant amount of liquid handling can also be avoided at this stage if you order premixed oligo sets, often done at no extra charge. In this case plates will be supplied with a known concentration of both forward and reverse primers mixed together in a single well. It should also be possible to obtain the remaining single primer plates left over from the mixing process although an additional fee may be charged for this.

### 3.2 Template Preparation

The following protocol describes tissue collection, using a dissecting microscope, from interaccession crosses that have been carried out 4 days prior to the tissue collection (*see* Fig. 1). The kits used for RNA purification, cDNA synthesis, DNase1 treatment, and quality control are listed in the materials section. However, workers should feel free to try their own combination of reagents.

1. Wipe down the work area, dissecting scope and utensils with RNaseZap, and confine subsequent sample and equipment processing to this zone.
2. Ensure an RNase-free 2 mL microcentrifuge tube is open and bend the lid far enough back to ensure it does not protrude over the upper rim of the tube. Place the tube on dry ice and cover with the upturned lid of a sterile plastic Petri dish (*see* **Note 4**).
3. Turn on the light source for the dissecting microscope.
4. Hold the stem of the silique with forceps and excise with a scalpel by slicing through the stem portion near to the plant.
5. Hold the silique in place under the dissecting scope with forceps and adjust the focus (*see* **Note 5**).
6. Gently draw the tip of a hypodermic needle along the length of the seed pod where one of the valves attaches to the septum.
7. Push open the valve wall to reveal the developing seed and if possible strip the valve wall away from the silique completely.
8. While holding the silique steady use the back of the needle to push each individual seed from its umbilical. Collect seed on the back of the needle and transfer to the frozen tube on dry ice.
9. If required gently tap the needle to dislodge the seed to facilitate their collection down into the tube.
10. Repeat the procedure until five siliques worth of seed have been collected.
11. Cap the tube and transfer to −70 °C freezer or keep seed on dry ice and proceed with RNA extraction.
12. Remove tube from dry ice and immediately pipette 450 μl of RNA extraction buffer into the tube.
13. Immerse tissue homogenizer (such as the *Omni International TH Tissue Homogenizer* with a *5 mm × 75 mm Flat Bottom SS Probe*) into the sample and thoroughly disrupt the tissues (*see* **Note 6**).
14. Proceed to purify RNA using the manufacturer's protocol.
15. Quantify RNA and check quality by microcapillary electrophoresis, e.g., Agilent Bioanalyzer (Agilent Technologies) or conventional fluorescent spectrophotometry and gel electrophoresis.
16. DNase1 treat 1 μg of total RNA and carry out cDNA synthesis using the manufacturer's instructions/standard laboratory protocols.

**Table 1**
**PCR program setup for PCR–HRM screening**

| *Setup* | | |
|---|---|---|
| Detection format | Block type | Reaction volume |
| SYBR Green I | 384 | 7 μL |
| *Programs* | | |
| Program name | Cycles | Analysis mode |
| Denature | 1 | None |
| Amplification | 50 | Quantification |
| HRM | 1 | Melting curve |
| Cooling | 1 | None |

### 3.3 Assay Setup and PCR Protocol

In our laboratory, HRM–PCR screening is carried out in 7 μL reaction volumes using a LightCycler 480 (Roche) in conjunction with the LightCycler 480 High-Resolution Melting Master reagents (Roche Cat. No. 04 909 631 001). Reagents were combined into 384-well plates using a CAS1200 liquid handling robot (Corbett Robotics). Final concentration of $MgCl_2$ was 2.5 mM and Primers were 0.2 μM. The amount of cDNA in a 7 μl reaction was approximately 2 ng. Any dilutions are made using PCR grade water (*see* **Note 7**).

It is recommended to have reactions arranged in blocks based on template type, i.e., all the false hybrid samples together, then the first parent samples, then the second parent, then interaccession direction one, interaccession direction two, and finally a water control. This will greatly facilitate entry of the assay details into the LightCycler software.

PCR conditions were designed to facilitate HRM assay of the maximum number of target genes conducive to a screening situation. PCR/HRM screening parameters and conditions are presented in Tables 1 and 2. As can be seen, these are split into four subprograms. We advise naming these as described in the tables, since the subsequent Gene Scanning module will link to the HRM subprogram by name later on. Since screening involves the use of many different sets of primers and primer design is constrained somewhat by positioning of the polymorphic sequence within the mRNA, it is advisable to use Touchdown PCR during the amplification stage. The Touchdown PCR described here has a first annealing cycle set at 65 °C and decreases by 0.5 °C in each subsequent cycle until it reaches 58 °C, whereby the annealing temperature remains constant for the remainder of the amplification. When using optimized HRM assays, a standard aim for amplification curve CP values is below 30 cycles.

**Table 2**
**Temperature parameters for PCR–HRM screening**

| Target °C | Acquisition mode | Hold (h:m:s) | Ramp rate °C/s | Second target °C | Step size °C | Step delay (cycles) | Acquisitions (per °C) |
|---|---|---|---|---|---|---|---|
| *Denature* | | | | | | | |
| 95 | None | 00:10:00 | 4.8 | | | | – |
| *Amplification* | | | | | | | |
| 95 | None | 00:00:10 | 4.8 | – | | | – |
| 65 | None | 00:00:10 | 2.5 | 58 | 0.5 | 1 | – |
| 72 | Single | 00:00:10 | 4.8 | – | | | – |
| *HRM* | | | | | | | |
| 95 | None | 00:01:00 | 4.8 | | | | – |
| 40 | None | 00:01:00 | 2.5 | | | | – |
| 65 | None | 00:00:01 | 1 | | | | – |
| 95 | Continuous | – | – | | | | 25 |
| *Cooling* | | | | | | | |
| 40 | None | 00:00:10 | 2.5 | | | | – |

However, for screening a range of genes on the same plate, we advise extending this to 45 cycles. This extended range is designed to accommodate not only variations in the transcript levels in the original RNA population but also variations in the first annealing cycle due to the touchdown strategy.

Once a run file has been created in the LightCycler software (LightCycler®480SW1.5), sample information can be added using the Sample Editor and Subset Editor screens accessible via the click button on the left of the screen. Sample blocks can be added in the Sample Editor, whereas individual subsets need to be entered for each gene of interest in the Subset Editor.

### 3.4 Data Analysis by Gene Scanning

These instructions assume the program detailed above has been successfully completed on the Roche LightCycler 480 and sample information has been input into to the Sample Editor and the Subset Editor modules.

1. Click on the "Analysis" button on the left of the software window.
2. In the "Create New Analysis" box, select "Gene Scanning" and a new window will appear.
3. In the "Subset" field, use the pull down options to select a gene identifier.
4. In the "Program" field, use the pull down options to select HRM.
5. Clicking on the icon button will take you to the HRM Gene Scanning Screen.

6. In the "Scanning Results" pull-down menu, select "Replicate of". This will color the curves based on the sample types entered during the program setup.
7. Click on the "Negatives" tab. Check on the amplification graphs to see if the water control is negative. If called negative by the software, it will not appear/feature in subsequent melting curve analysis.
8. Click on the "Normalization" tab. Use the sliders on the upper melting curve graph to select a single melting domain from the curves. The Green sliders should be on the high side of the melt (left) and the blue sliders on the low (right side). The area within each slider creates a reference section of the trace where the curves will be pulled together. Adjusting the position and width of the slider, will delineate the section and to some degree morphology of the melting curves used for the normalization. Moving the sliders automatically updates the main Normalized Melting Curve chart, which is also visible on this screen. At this stage, an obvious difference should be evident between the False Hybrid trace and both of the pure parental curves.
9. Click on the "Temp Shift" tab.
10. Use the up and down arrows on the Threshold window until the Normalized and Shifted Melt Curves gather together and overlay each other at the base of the curves (usually a value of approximately 5 units will suffice).
11. Click on the "Difference Plot" tab.
12. Click on the "Calculate" button and a difference graph should appear in the blank plot area.
13. Click on the "Select Base Curve" button, select the False Hybrid on the pop-up screen and click the icon button to apply. The resulting graph should be visually assessed to provide evidence of monoallelic expression. If monoallelic, the interaccesion curves will match the morphology of the parental curves very closely. The parental curves should deviate from the straight false hybrid reference line in an obvious manner (*see* Figs. 2 and 3).
14. To output the graph or the raw plot data, right click on the chart and select the "Export Chart" option from the pop-up menu. Select the Picture or the Data tab and enter the image format required (if appropriate) and/or enter a file name and save destination.
15. When appropriate saves have been selected click on the 🖫 icon to save the data.
16. Repeat the PCR run and analysis on at least two further biological replicate samples for each gene assayed and/or validate using an alternative method (*see* **Notes 8–10**).

## 4 Notes

1. During PCR for HRM, *specificity*, and not *sensitivity*, is the main aim. Thus, relatively low primer concentrations are used and the lowest $MgCl_2$ concentration that enables amplification is encouraged.
2. HRM can be separated from the PCR step if required. For example, PCR products can be generated on a basic PCR machine using normal reagents. Fluorescent dye can then be added to the PCR tubes and melting analysis carried out on a machine capable of HRM. This need not be a PCR machine and stand-alone HRM platforms are becoming available.
3. When using these criteria, for approximately 300 arabidopsis genes, we designed amplicons that ranged in size from 62–339 bp (average length 158 bp). For 25 % of the loci, no suitable intron was present in the vicinity of the SNP and/or no suitable primers could be designed over an intron–exon junction.
4. This is to stop ice crystals forming on the inside the tube during tissue dissection.
5. Some workers find it hard to hold the silique in place and so attach the silique to a glass slide with double-sided sticky tape.
6. We found that use of the TH homogenizer increased our total RNA yield approximately sevenfold compared to manual disruption using a micropestle.
7. Reagent dispensing and mixing for large-scale PCR-based screens can take a long time and most liquid handling robots are not cooled. Therefore, we advise the use of a Taq DNA polymerase modified to show no activity at room temperature (like the FastStart Taq included in the Roche HRM Master). For example, FastStart Taq DNA polymerase activity is blocked and shows no activity up to 75 °C. This stops reductions in specificity and sensitivity due to nonspecific amplification products. If possible we also suggest that tube/plate holders from the robot are prechilled in the fridge prior to the mixing run.
8. If it is desirable to develop an assay that confirms which direction the parental bias is present just from the melting curves, but low melt differences are apparent between the pure parental samples, it may be possible to spike the samples. Spiking all the samples 1/10–1/5 with one parental cDNA may introduce enough heterodimer to accentuate differences between the parental populations while still maintaining difference to the False Hybrid.
9. We note the melting behavior of the interaccession crosses and repeat the whole experiment with a duplicate biological repeat.

We then perform further PCR on other biological material, purify the PCR product, and then have it Sanger sequenced using one of the original PCR primers. It is also possible to sequence the PCR products from the real-time PCR/HRM experiment *en masse*. This can be done by combining amplicons by template, column purifying the resultant mixture to remove primers and dye, etc., and having adaptor-ligation libraries made for FLX sequencing.

10. This screening protocol relies on large melting differences. As such it is worthy of note that we have also been able to confirm monoallelic expression using the conventional non-saturating SYBR real-time PCR dye.

## Acknowledgment

Protocols were developed in the Macknight laboratory with support from the Marsden Fund of New Zealand.

### References

1. Berger F, Chaudhury A (2009) Parental memories shape seeds. Trends Plant Sci 14: 550–556
2. Scott RJ, Spielman M, Bailey J, Dickinson HG (1998) Parent-of-origin effects on seed development in *Arabidopsis thaliana*. Development 125:3329–3341
3. Gehring M, Bubb KL, Henikoff S (2009) Extensive demethylation of repetitive elements during seed development underlies gene imprinting. Science 324:1447–1451
4. Gehring M, Missirian V, Henikoff S (2011) Genomic analysis of parent-of-origin allelic expression in *Arabidopsis thaliana* seeds. PLoS One 6:e23687
5. Hsieh TF, Shin J, Uzawa R, Silva P, Cohen S, Bauer MJ, Hashimoto M, Kirkbride RC, Harada JJ, Zilberman D, Fischer RL (2011) Regulation of imprinted gene expression in Arabidopsis endosperm. Proc Natl Acad Sci U S A 108:1755–1762
6. McKeown PC, Laouielle-Duprat S, Prins P, Wolff P, Schmid MW, Donoghue MTA, Fort A, Duszynska D, Comte A, Lao NT, Wennblom TJ, Smant G, Köhler C, Grossniklaus U, Spillane C (2011) Identification of imprinted genes subject to parent-of-origin specific expression in *Arabidopsis thaliana* seeds. BMC Plant Biol 11:e113
7. Shirzadi R, Andersen ED, Bjerkan KN, Gloeckle BM, Heese M, Ungru A, Winge P, Koncz C, Aalen RB, Schnittger A, Grini PE (2011) Genome-wide transcript profiling of endosperm without paternal contribution identifies parent-of-origin-dependent regulation of AGAMOUS-LIKE36. PLoS Genet 7: e1001303
8. Wolff P, Weinhofer I, Seguin J, Roszak P, Beisel C, Donoghue MTA, Spillane C, Nordborg M, Rehmsmeier M, Köhler C (2011) High-resolution analysis of parent-of-origin allelic expression in the Arabidopsis endosperm. PLoS Genet 7:e1002126
9. Luo M, Taylor JM, Spriggs A, Zhang HY, Wu XJ, Russell S, Singh M, Koltunow A (2011) A genome-wide survey of imprinted genes in rice seeds reveals imprinting primarily occurs in the endosperm. PLoS Genet 7:e1002125
10. Weigel D, Mott R (2009) The 1001 genomes project for Arabidopsis thaliana. Genome Biol 10:e107
11. Day RC, Herridge RP, Ambrose BA, Macknight RC (2008) Transcriptome analysis of proliferating Arabidopsis endosperm reveals biological implications for the control of syncytial division, cytokinin signaling, and gene expression regulation. Plant Physiol 148:1964–1984
12. Day RC, McNoe L, Macknight RC (2007) Evaluation of global RNA amplification and its use for high-throughput transcript analysis of laser-microdissected endosperm. Int J Plant Genomics 2007:e61028
13. Wittwer CT (2009) High-resolution DNA melting analysis: advancements and limitations. Hum Mutat 30:857–859
14. Rozen S, Skaletsky H (2000) Primer3 on the WWW for general users and for biologist programmers. Methods Mol Biol 132:365–386

Chapter 6

# Analysis of Genomic Imprinting by Quantitative Allele-Specific Expression by Pyrosequencing®

Peter C. McKeown, Antoine Fort, and Charles Spillane

## Abstract

Genomic imprinting is a parent-of-origin phenomenon whereby gene expression is restricted to the allele inherited from either the maternal or paternal parent. It has been described from flowering plants and eutherian mammals and may have evolved due to parental conflicts over resource allocation. In mammals, imprinted genes are responsible for ensuring correct rates of embryo development and for preventing parthenogenesis. The molecular basis of imprinting depends upon the presence of differential epigenetic marks on the alleles inherited from each parent, although in plants the exact mechanisms that control imprinting are still unclear in many cases. Recent studies have identified large numbers of candidate imprinted genes from *Arabidopsis thaliana* and other plants (see Chap. 7 by Köhler and colleagues elsewhere in this volume) providing the tools for more thorough investigation into how imprinted gene networks (IGNs) are regulated. Analysis of genomic imprinting in animals has revealed important information on how IGNs are regulated during development, which often involves intermediate levels of imprinting. In some instances, small but significant changes in the degree of parental bias in gene expression have been linked to developmental traits, livestock phenotypes, and human disease. As some of the imprinted genes recently reported from plants show differential rather than complete (binary) imprinting, there is a clear need for tools that can quantify the degree of allelic expression bias occurring at a transcribed locus. In this chapter, we describe the use of Quantification of Allele-Specific Expression by Pyrosequencing® (QUASEP) as a tool suitable for this challenge. We describe in detail the factors which ensure that a Pyrosequencing® assay will be suitable for giving robust QUASEP and the problems which may be encountered during the study of imprinted genes by Pyrosequencing®, with particular reference to our work in *A. thaliana* and in cattle. We also discuss some considerations with respect to the statistical analysis of the resulting data. Finally, we provide a brief overview of the future possibility of adapting Pyrosequencing® for analyzing other aspects of imprinting including the analysis of methylated regions.

**Key words** Imprinting, Pyrosequencing®, QUASEP, Parental conflict, *Arabidopsis*, Cattle

## 1 Introduction

### 1.1 Genomic Imprinting in Plants

Genomic imprinting describes a parent-of-origin phenomenon in which the alleles of a gene are expressed in a parentally biased manner [1]. This causes the allele inherited from one parent to be preferentially or exclusively expressed in comparison to that inherited from the other. Imprinted genes can either be imprinted

Charles Spillane and Peter C. McKeown (eds.), *Plant Epigenetics and Epigenomics: Methods and Protocols*, Methods in Molecular Biology, vol. 1112, DOI 10.1007/978-1-62703-773-0_6, 

maternally expressed genes (iMEGs) or imprinted paternally expressed genes (iPEGS). Imprinting occurs due to differences in the epigenetic marks associated with the two alleles that are established during either make or female gametogenesis although the mechanisms are not always clear.

Despite the fact that genomic imprinting was first discovered in flowering plants (Angiosperms; see McKeown and Spillane, Introduction to this volume, Chap. 1), it is better known through its regulation of large numbers of genes in eutherian mammals (reviewed [2]). Recent work has increased the number of candidate imprinted genes in three plant species, *Arabidopsis thaliana, Oryza sativa*, and *Zea mays*, which indicates that imprinting is likely to regulate several hundred genes in each species [3]. The first imprinted genes were identified on the basis of the non-Mendelian inheritance of their seed phenotypes, which identified *MEDEA* and *FIS2* as iMEGs. MEDEA and FIS2 are components of the FIS complex, which regulates the rate of endosperm development and was also found to restrict expression of *PHERES1* [4]. *PHE1* was the first iPEG to be identified. *OsFIE* was subsequently identified as a member of the rice homologue of the FIS complex and *AGL62* as a possible homologue of *PHE1* [5], which extended the analysis of genomic imprinting in plants to a third plant species.

Although fewer than 20 plant genes were known to be imprinted either as iMEGs or iPEGs as of 2009, it could be expected that the actual number in any given Angiosperm species would be an order of magnitude greater, in line with the extent of imprinting in mammalian systems. Candidate imprinted genes were proposed from *Arabidopsis* on the basis of endosperm-specific expression [6], links to DMRs [7, 8] or similarity to the known imprinted gene *FWA* [7]. The extent of conservation of imprinting in *MEDEA*-related genes from other species was also studied [9, 10]. Recently, a combination of RNA-seq, mutant modifier and cDNA-AFLP approaches allowed several hundred plant genes to be proposed to be regulated by genomic imprinting. These include studies performed in *Arabidopsis thaliana* [11–14], rice [15], and maize [16, 17] and may allow further analysis of the evolutionary forces, which have shaped genomic imprinting in plants (see below; [10, 18]).

The identification of large numbers of imprinted genes in multiple systems suggests the possibility that these can be used to critically assess theories for explaining the evolution of imprinting. Prominent amongst these are the parental conflict theory (or PCT) [19, 20] and modern adaptations of this related to degrees of outbreeding [21]. A prediction of this theory is that approximately equal numbers of imprinted genes are expected to be maternally and paternally expressed, but at present more maternally expressed imprinted genes are known from plants. This could be due to low expression of PEGs, and in *Arabidopsis thaliana*, to a failure to detect iPEGs due to contamination from the maternally derived seed coat (see below). This also suggests that there is a particular need for the use of quantitative techniques to distinguish between this and other theories for the evolution of imprinting, such as dosage control and maternal coevolution [22, 23], and to assess the level of differential vs. binary imprinting. For a discussion of the use of high-resolution melting (HRM) analysis to detect lowly expressed imprinted genes from hybrid plant seed cDNA see Chap. 5 of Day and Macknight in this volume. Here, we consider the robust technique Quantification of Allele-Specific Expression by Pyrosequencing® (QUASEP) as a means of quantifying the extent of uniparental expression of imprinted genes (or indeed any genes displaying uniparental expression patterns).

### 1.2 Pyrosequencing® as an Allele-Quantifying Technology

Pyrosequencing® is a "sequencing by synthesis" technology, which utilizes the release of pyrophosphate during DNA synthesis to synthesize ATP, which is used by the luciferase to release a fluorescent signal. This fluorescence is then measured and the identity of the nascent sequence determined in real-time. This chain reaction works on a stoichiometric manner, as one molecule of ATP is created during one base incorporation, releasing one "unit of light" by the luciferase, allowing the quantitative detection of SNPs during the sequencing reaction. Pyrosequencing® displays many features useful for DNA sequencing, notably the fact that the fluorescent chemistry, which it employs is highly quantitative. It also permits a significant degree of automation [24]. Of particular value is the fact that it "dispenses with the need for labeled primers, labeled nucleotides, and gel-electrophoresis" [25, 26]. As such, Pyrosequencing® chemistry forms the basis for many sequencing platforms, including 454, and a historical overview of its development can be found in the opening paragraphs of [27]. However, for the purposes of this review, we concentrate on the use of Pyrosequencing® in the characterization of imprinted genes. More specifically, we discuss the use of Pyrosequencing® in the context of QUASEP, which allows highly accurate quantification of the relative proportions of bases at a defined SNP within a mixture of genomic DNA or PCR products. The QUASEP technique has been widely used in studies of imprinting in animal systems and could be of use to the plant imprinting community for providing detailed studies of imprinted genes in the future. It should be noted that the protocols that we describe can also be applied to analysis of any system in which different alleles of a heterozygous gene are expressed, as well as indel analysis, SSRs, copy number variation, and others: many of the issues associated with these applications have been dealt with in detail in previous entries in this series [28]. We therefore concentrate on potential problems relating to the study of imprinting in plants. Finally, we consider some possible future directions for the application of QUASEP to our understanding of imprinting. These include the study of differentially methylated regions (DMRs), which act as imprinting control regions (ICRs), with reference to protocols which are already well established in animal systems; and second, the possibility of using Pyrosequencing® to identify de novo SNPs for analysis of imprinted genes in less well-studied species.

### 1.3 Applications of Pyrosequencing® to Analysis of Imprinted Loci

The ability of Pyrosequencing® to accurately quantify the level of allelic imbalance is clearly suitable for the detailed analysis of candidate imprinted genes. Before turning to a discussion of a typical QUASEP workflow as applied to the study of imprinted genes, we consider more specifically how Pyrosequencing® can assist in the identification of imprinted genes, and their use in critical assessments of theories for the evolution of imprinting. The identification of imprinted genes depends upon two parallel lines of evidence: the gene must be uniparentally expressed under given conditions, and this must also be shown to be due to epigenetic marks leading to monoallelic expression [29]. Hence, it is necessary to distinguish imprinting from gamete deposition effects and allelic dominance. A further complication, which is particularly important in *A. thaliana*, is that imprinting is restricted to the endosperm, the terminally differentiated F1 tissue, which nourishes the embryo during its development, even in cases in which the gene is expressed in other tissues [30]. For some genes, it may therefore be necessary to distinguish between expression from the maternally derived seed coat, biallelically expressed embryo tissue, and the endosperm in which imprinting may occur. (This is effectively the plant equivalent of placental contamination in studies of imprinting in mammalian fetuses, as discussed by [29].)

Quantitative analysis of imprinting is especially important in cases where the degree of imprinting varies between different tissues. Two cases can be distinguished in such instances: cases in which a gene is expressed in a uniparental manner in one tissue but in a biallelic manner in others; and those in which a gene is imprinted in multiple tissues, but to a different degree. Additionally, instances are known in which the *direction* of imprinting differs in different tissues, e.g., a gene which is maternally expressed in one tissue but paternally in another. Instances of this are better known in mammalian systems than in plants, but may come to be of greater utility now that more imprinted genes have been identified. For example, in animals, the *GNAS* gene is expressed from most human tissues but had been described to be imprinted only in the pituitary, thyroid, renal proximal tubules, and gonads [31]. However, careful use of controlled Pyrosequencing® experiments, followed by statistical analysis, determined that expression in four further tissues was also significantly maternally biased ($p < 0.0001$) by between 54.1 and 56.1 % depending on the tissue [32]. The authors of this study suggested that this may represent a general trend towards maternal expression, which would have been obscured had an "on/off" model of imprinting been followed.

The major limits to such applications concern the need to generate suitably pure material. In the case of plants, the most significant tissue for studies of imprinting is the endosperm, but dissection of endosperms from seed is complex and prone to contamination. There is significant evidence that different imprinted genes are particularly expressed from certain regions of the endosperm, suggesting that quantification by Pyrosequencing® may need to be applied to isolated endosperms, which have themselves been dissected in order to provide a clear picture of the interactions between imprinted genes in different areas of the seed. These challenges also have a temporal dimension as younger seeds are smaller and harder to dissect accurately. For descriptions of the best approaches to isolation of pure endosperm by LCM, see [33, 34] and references therein. Angiosperm endosperm can also be distinguished on the basis of developmental stage or different fraction (chalazal, peripheral, micropylar: for approaches to manually isolate such fractions, see ref. 35). For highly accurate dissection of regulatory effects on plant imprinting, this may be significant, especially as imprinting may be preferentially associated with the chalazal fraction of the endosperm [14].

### *1.4 Strategies for Identifying Partial Imprinting*

The techniques, which we describe in this chapter, have so far only been applied within the plant imprinting field to quantification of SNPs as a means of describing imprinted genes. However, the main advantage of QUASEP is that it allows an accurate analysis of partially monoallelic gene expression, e.g., due to partial imprinting. This may arise from at least two sets of circumstances. For example, a locus may be imprinted in certain genetic backgrounds but not in others [14] as was widespread amongst PEGs identified by Gehring and colleagues [11]. Such loci can be identified by nonquantitative Sanger sequencing, as it is only necessary to distinguish presence from absence in each case. We have previously identified an example of a different kind, in the case of the *MS5* gene, which not only is preferentially maternally expressed in the endosperm but also displays some seed coat expression [13]. In a case such as this, performing QUASEP on very pure tissue samples, such as those derived by LCMD, could clarify the degree of uniparental expression and the factors involved in regulating it. Proof of principle for these may be easier to derive in cereals (maize and rice), in which seeds are large enough to be manually dissected without significant risk of contamination. Even in these cases, at least one partially imprinted gene is known from maize [13], indicating that the phenomenon of partial imprinting is not an artifact caused by seed coat contamination.

In our work on cattle, we have shown that relative levels of polymorphic SNPs in both coding and noncoding regions of the imprinted *GNAS*, *IGF2*, and *DLK1-DIO3* domains can be correlated with phenotypes of interest, including agricultural traits [36–38]. We were not able to determine from these studies whether such SNPs are causal for phenotypes, as may be the case for one SNP which introduces a nonsynonymous substitution, but the relative proportions of noncoding SNPs may be of interest as markers for particular traits. Hence, SNPs in the imprinted genes of crop species could be of value as markers in breeding programs if it were possible to find the expected associations between such imprinted genes and seed phenotypes.

## 2 Materials

### 2.1 PCR Reagents

1. cDNA. For the purposes of this chapter, it will be assumed that cDNA has been generated following any suitable RNA extraction and cDNA synthesis and has been confirmed to be of suitable concentration and free of any DNA contamination. Reagents associated with these steps are not therefore listed.
2. Primers, biotinylated. *See* **Note 1** for description of issues with suppliers.

   One biotinylated primer, one reverse primer (typically generating 150–200 bp amplicon), and one sequencing primer, *see* Subheading 3.2 for details.
3. Ampligold Taq system or similar. Applied Biosystems (but *see* **Note 2**).
4. Autoclaved MilliQ water ($ddH_2O$).
5. Electrophoresis grade Agarose. Bioline, Catalogue #41025.
6. SYBR safe nucleic acid gel stain (replaces ethidium bromide). Invitrogen, Catalogue #S33102.
7. 100 bp ladder.
8. 1× Tris-buffered EDTA. Sigma Aldrich, Catalogue ## E5134, T3253.

### 2.2 Pyrosequencing Reaction

1. PSQ reaction buffers and reagents (enzyme and substrate). Qiagen, Catalogue #970802 for Q24, #972812 for Q96. Includes Binding Buffer (10 mM Tris–HCl, 2 M NaCl, 1 mM EDTA, 0.1 % Tween 20 (vol/vol) at pH 7.6), Annealing Buffer (20 mM Tris-acetate, 2 mM Mg-acetate at pH 7.6) and Wash Buffer (10 mM Tris-acetate at pH 7.6). To be kept at 4 °C.
2. Streptavidin Sepharose beads, preserved in 20 % ethanol. GE Healthcare, Catalogue #17-5113-01. To be kept at 4 °C.
3. 0.2 M NaOH. Sigma, Catalogue #221465.
4. 70 % ethanol. Sigma-Aldrich, Catalogue #E7033 or any suitable supplier.
5. Autoclaved MilliQ water ($ddH_2O$).

6. Pyromark cartridge. Qiagen, Catalogue #979202 for Q24 or #979004 for Q96.
7. USB stick to transfer assay to PSQ machine and retrieve completed data.

## 3 Methods

The protocol as developed here originally derives from techniques developed for the application of QUASEP to SNP analysis [39], especially with respect to the sequencing reaction itself. There are few safety issues to consider beyond those relating to good practice for standard molecular biology, although care should be taken with the use of 0.2 M sodium hydroxide (irritant) and 70 % ethanol (flammable), and with the use of the hot-plate. Clean disposable gloves and other suitable protective equipment should be worn throughout. Maintenance of Pyrosequencing® machines should be performed according to the manufacturer's guidance (see e.g., http://www.qiagen.com/products/pyromarkq96idandpyromarkq96mdaccessories.aspx #Tabs=t1) and is not discussed further here except to note that both PSQ machines and the associated wash stations ought to be placed on benches of a suitable width and height, be kept horizontal and kept covered when not in use. The head of the wash station adapter is inclined to accumulate dust even after short periods of inactivity so should be covered with a suitable plastic sheet and cleaned often. For specific advice relating to the maintenance of the PSQ cartridges, *see* Subheading 3.5 below.

### *3.1 Identification of Probable Candidate Genes*

A particular role for Pyrosequencing® is in the validation of genes predicted to be imprinted on the basis of next-generation sequencing screens. This, for example, was how Wang and colleagues confirmed 17/26 genes predicted to be imprinted by Illumina/Solexa sequencing of mouse neonatal brain tissue, thus validating their sequencing experiment [40]. Only three of those genes were however novel, and the use of Pyrosequencing® is more commonly used for the more detailed analysis of known imprinted genes in different tissues or under different conditions. For simple validation purposes, it may typically be cheaper to use normal Sanger sequencing of PCR products spanning an SNP, as this avoids the cost of QUASEP reagents and the purchase of a biotinylated primer. The use of CAPS-PCR, a semiquantitative technique [41], is also possible. An alternative approach is to use QUASEP to confirm genes predicted to be imprinted by bioinformatic screens [42]. Depending on the precise algorithms used, this again prove more expensive as the false-discovery rate is likely to be elevated (only 2 out of 16 predicted instances of imprinting were confirmed by Ruf and colleagues, for example, even allowing for partial and strain-specific effects [43]). Uniparental inheritance can however be established by various techniques, such as the multiple lines of evidence used for identifying porcine imprinted genes [44] in which gene expression in parthenogenetic fetuses was combined with Pyrosequencing® analysis of fetal tissue from parents of different strains. Such a combinatorial approach has the advantage of ensuring that artifacts are not introduced by hybridization between pigs of two genotypes. Issues relating to the identification of candidate imprinted genes from existing datasets are discussed in more detail in the chapter by Korir and Seoighe in this volume (Chap. 4).

### 3.2 Assay Design Considerations

The principal issue is to design primers, which are capable of amplifying a very specific area, as not all genes are SNP-rich. As will be explained in Subheading 3.3, the key determinant of a successful assay is its ability to generate a specific product. Although entirely erroneous products can be readily distinguished by visualization of a gel, care must also be taken not to amplify regions of genes highly similar to the candidate. For example, the *A. thaliana* imprinted gene *PHERES1* has high similarity to its nonimprinted neighbor, *PHERES2*, which shares its endosperm expression pattern (Genechip; [35]), so primers that distinguish the two are needed in order to analyze their expression by QUASEP [13]. Particular care would also be needed in the analysis of imprinted genes, which belong to large families, especially the *AGAMOUS-LIKE* family of TFs, of which multiple members are imprinted in both monocots and dicots [14, 15, 45]. For issues related to assay design, *see* **Note 1**.

Whichever design procedure is chosen, it is essential that one of the primers is designed to contain a 5′-biotininylated end which at present is the only modification compatible with the Pyrosequencing® chemistry. The assay(s) is then assigned to particular wells in the Run Design software, together with sample names and all other relevant information. This generates the "prerun information" which includes the volumes of Enzyme and Substrate, and of each dNTP, which will be needed to complete the run. The software settings, which are initially provided or chosen as defaults, do not always correspond to the parameters of the cartridge and Pyrosequencing® machine being used so it may be necessary to install and select them manually (the parameters relevant to a particular make of cartridge, which relate to the rate of dispensation, are displayed on the label on its side).

### 3.3 Initial and Nested PCR to Generate Biotinylated Products

For running the QUASEP, the region surrounding the SNP of interest is amplified from the prepared cDNA by PCR. We have used an initial denaturation step of 5 min at 95 °C, followed by 50 cycles of an amplification program of 30 s at 95 °C for denaturation, 30 s at the appropriate primers annealing temperature and 30 s at 72 °C for extension. The final extension step was performed after 50 cycles at 72 °C for 5 min (*see* **Note 2**). Importantly, PCR products were always checked on gels prior to the performance of the Pyrosequencing® reaction (these were typically 1.5 % w/v agarose gels stained with any safe system in 1× TBE buffer). In order to obtain reliable quantitation results, a strong PCR band is required, as low amounts of PCR fragments will result in low-intensity/low-quality peaks during the pyrosequencing reaction. While using the PSQ96 system, these reactions were performed in 50 μl volumes where each contained 20 ng of DNA or cDNA, 0.6 μM of each primer, GeneAmp 10× PCR buffer, 10 mM of each of dNTP, 2.5 mM $MgCl_2$, and 1.25 μl of AmpliGold Taq (Applied Biosystems) but *see* **Note 2**. PCR products can be stored at −20 °C at this point, at least for a limited period, although biotinylated products are light sensitive so sealed containers may be recommended. Particular care should also be taken to avoid exposing such primers to multiple freeze–thaw cycles. Some protocols recommend the use of a second round of nested PCR, for a discussion of which *see* **Note 3**. PCR should also be performed on suitable controls (*see* **Note 4**).

### 3.4 Sample Preparation

The remaining PCR is mixed with an equal volume of binding buffer master mix containing nine parts Binding Buffer to one part Streptavidin-Sepharose™ beads (GE Healthcare, Amersham, Little Chalfont, UK). *See* also **Note 5**.

The mixture of the product and beads is then shaken at 1,400 rpm at room temperature for 5–10 min. The use of this high speed is essential for generating a vortex that will ensure full mixing of the biotinylated PCR product to the beads. We simply used masking tape to stick the plate to a normal mixer such as an Eppendorf Thermomixer Compact (Eppendorf). While the products are shaking, the wash station should be prepared. This should be positioned as near to the shaker to ensure no settling occurs, as well as to the Prosequencing machine. The wells of the station should be filled with 70 % v/v molecular grade ethanol, denaturation solution (0.2 M NaOH, which should be freshly made but need not be autoclaved) and 1× Wash Buffer, in the positions indicated by the manufacturers. The head of the Vacuum Prep Tool is prepared by washing it through with MilliQ water. The PSQ HS 96- or 24-well plate should also be prepared at this point and will contain either 40 or 25 μl of annealing solution, respectively (1.6 or 1.0 μl of sequencing primer made into 10 pmol aliquots, and 38.4 or 24 μl of Annealing Buffer or any other combination to give a final primer concentration of 10 μM).

The streptavidin–biotin-bound PCR products are then captured with the probes of the Pyrosequencing® Vacuum Prep Tool (Biotage, Charlottesville, VA), and the adhered products washed in each solution—ethanol, NaOH, and wash buffer—for 12 s for each step. The vacuum is then turned off and the beads are then released into the plate containing the annealing solution—this step easily introduces contamination if the probes are not inserted into the correct wells.

Attention must also be paid to ensure that there is no splashing between the wells, which are very shallow, and that the vacuum has been switched off prior to putting the probes in, to avoid any residual suction of the annealing buffer and incorrect release of the beads! The product is then incubated for 2 min at 80 °C on a heating block (an open PCR machine works perfectly well for this) to anneal the sequencing primer to the single-stranded PCR product. While this is happening it is appropriate to wash the vacuum head through with water two more times before switching off the pump.

Finally, the plate is allowed to cool to room temperature (for approximately 5 min), while the cartridge is prepared by adding the requisite volumes of enzyme, substrate and dNTPs, as determined by the Pyrosequencing software. The enzyme and substrate should be freshly resuspended and used within a week (if kept at 4 °C), so it makes financial sense to arrange to perform several runs in this period. This is particularly the case for the Q24 as 3–4 plates can typically be run from a single pair of lyophilized reactives. We have also used aliquots of enzyme and substrate stored at −20 °C for up to 3 weeks although this is not recommended by the manufacturers. For both the reactives and the dNTPs, the requisite volumes should be pipetted down the side of the cartridge well and the introduction of bubbles avoided, although we have never been aware of any problems from this stage. The plate and cartridge are inserted into the Pyrosequencing® machine [for us, the PSQ HS or Q24 Pyrosequencer (Qiagen)] alongside the PSQ Cartridge and the experiment run. The data are generated in real time and can then be analyzed (*see* **Notes 6** and 7).

### 3.5 Clean-Up

After completion of a run, the materials are discarded and the wash station cleaned. The cartridge should be washed through with $ddH_2O$, then stored for reuse. *See* **Note 8**.

## 4 Notes

Note 1: Assay design

Although we have designed PCR primers spanning coding SNPs of interest using the provided Biotage PSQ Assay Design Software, this is only essential for providing the assay with the predicted products of the sequencing reaction during the Pyrosequencing® run. Hence, any other program can be used to design the primers for the initial PCR, and this may be preferable for some applications, provided that the maximum recommended read length (70–100 bp) is observed. The Assay Design software does not allow primers to be designed to a particular region, or to avoid another which is known to have high similarity to another locus in the relevant genome, for example. Nor does it permit the enforcement of a GC clamp. A category of problem which is harder to deal with, because it is less easily explained, relates to issues with the primers themselves and especially to the sequencing primer. The Assay Design software assigns confidence scores and subsequent rankings to assays, even those which have high scores and are classed as optimal may give poor peaks or strongly biased reads. Contamination with related PCR products may explain such problems in some cases, but we are also aware of users encountering such issues when using preprepared "test DNA" designed for validation of QUASEP assays for bisulfite-treated DNA. In some cases, these problems were linked to the choice of the manufacturer for either the biotinylated primer or even the sequencing primer—how this latter could cause differences is unclear although we have found Biotez to be overall the most reliable (http://www.biotez.de/index.php/en/oligonucleotides.html).

Note 2

For the PCR to generate the biotinylated amplicon, the use of a large number of cycles of short duration has been found to be optimal for generation of highly specific products of small size. Initially, we performed PCR using the AmpliGold system to ensure product specificity, although any proofreading enzyme should be suitable. A significant advantage of the reagents as currently supported, and which we have used with the PSQ24, is that far smaller volumes of PCR product are needed, 10–20 μl of product, which has at least halved the cost in terms of PCR reagents. We now typically perform PCR using ready-prepared master mixes, which often include gel buffers—as the PCR product will effectively be purified during the subsequent vacuum washing stage, this poses no problem for the Pyrosequencing® reaction.

Note 3

It is our experience that nested PCR design is not always critical, as the overwhelming determinant of successful Pyrosequencing® is specificity. Hence, adjustment of the reaction conditions to ensure that a single band is obtained with minimal primer–dimer formation typically obviates the need for a second round of PCR even if the product appears relatively weak. A nested PCR design can however be of use where persistent specificity issues are encountered [46]. Most imprinted protein-coding genes of sequenced model organisms have been found to contain suitable regions for use as unique templates, although analysis of smaller ncRNA genes, which have also been predicted to be maternally expressed in plants [47], might be expected to be more challenging. Nested PCR could be suitable in the case of RNA genes for this reason. Use of bisulfite-treated DNA to analyze DMRs associated with imprinted loci also suffers from this issue (see below—reviews of designing primers for bisulfite-treated regions should be followed in this instance). We have not experimented with gel extraction as a means of ensuring specificity although theoretically this should also be possible.

Note 4: Controls and replicates

It is crucially important to use appropriate controls for each assay. As has previously been emphasized, it is essential to use gDNA to detect any skew in base detection due to PCR bias [28, 29]. If there is a deviation from the 100 % X–50 % X/50 % Y–100 % Y series expected for the two parents with an XY SNP within the amplicon, then results from seed samples should be normalized accordingly. However, it would seem sensible to suggest that if the error is greater than, say, 5 %, a new assay should be designed. Such controls will also exclude incorrect conclusions due to DNA polymorphisms [29], which may be relatively widespread in out-crossing species. It should be noted that for detection of imprinting in seeds, the gDNA for these controls should be derived from adult tissue of the relevant plants, as hybrid seed will contain an unknown mixture of the maternal and offspring genomes. The control should also be performed for both parental genotypes, as the most common situation is for an assay to have a bias towards one allele (overcalling A as opposed to G, for example). In the case of partial imprinting, gDNA controls can also be used to generate standard curves by analyzed mixtures containing different predefined ratios of each parental allele. Although this has not been applied to the analysis of imprinting itself, it is usual in the case of partially methylated DMRs and has also been used to accurately determine the levels of SNPs that lead to alternative splicing events [48]. The development of such standard curves could therefore enhance the sensitivity with which degrees of partial imprinting can be analyzed in a similar manner and aid in the detection of small but significant effects of modifier phenotypes.

Note 5

The usual care should be taken to resuspend sepharose beads thoroughly before use, to keep them refrigerated and to use them within their expiry limit. This stage can conveniently be performed with a multichannel pipette and a BB master mix made up in a sterile plastic trough, in which case excess BB master mix should be made.

Note 6: Interpretation of QUASEP data

Data from completed runs is read using the analysis software version 1.2 in Allele Quantification mode. (It is at this stage that results will be categorized as "Successful," "Check," or "Failed," the reasons for which are considered below.) Assuming that the runs have worked successfully, the data from the plate should be analyzed with respect to the controls. For a new assay, four controls are recommended, to test the biotinylated primer without sequencing primer, the sequencing primer without PCR product, the sequencing primer with water, and water alone. For the smaller plates provided with the Q24, these typically account for a single plate, and the first four need not be repeated.

Note 7: Failed runs due to poor sequencing traces

Failed runs are automatically called for wells in which the inputted reference sequence is not detected, so if an unannotated SNP is present a run with excellent peaks may be called as a "Fail." For this reason, each run should be analyzed individually. If a misannotation in the environs is suspected, then the region can be submitted for Sanger sequencing and the reference sequence manually altered. Other than this, the main reasons for a run failing or being marked as a "Check" are that the sequencing peaks do not meet quality criteria. Again, each well should be considered individually as "Check" wells may be suitable for use if the SNP is correctly sequenced. Contamination between wells should be avoided carefully, which the use of empty controls can control for. Reactions which fail to meet quality criteria are usually easily diagnosed, most commonly due to a degraded, nonspecific or low concentration product. If the product is present at a low concentration, various technical issues may be responsible. Issues have previously been raised with respect to the design of the vacuum attachment, which could be purchased with the wash station [46]. In this report, the authors took to deconstructing the vacuum head and building their own version along more suitable lines, although they admitted that such a feat might be beyond the budget or expertise of many labs. The principal issues with the original vacuum head related to the fact that it consisted of two small chambers with a thick and unnecessary barrier between them and only one outlet. This led to uneven pressure, particularly in the probes at the far end of the vacuum head. We ourselves observed occasional blockages, typically in the terminal probes.

Note 8: Cleaning and storage

If the wash station will not be used again for some time, it is appropriate to cover it with plastic as we have found that it gathers dust very quickly. The most important thing after a run is to flush through the cartridge with MilliQ water or similar, filling the wells and then applying pressure with a gloved finger to expel the water through the dispensation needles in the base. This should be done several times. If the needles are in working order, the water will issue as a continuous, straight stream. If it is expelled at a skewed angle, or much slower than expected, it is likely that the needle is blocked or damaged and the cartridge's function will be impaired. Cartridges can be repeatedly used as long as the needles are operative. After cleaning, a cartridge should be left to air-dry and then stored in a suitable box. It is neither desirable to dry the cartridge with paper towels nor to leave them stored in paper, as the needles are easily damaged by stray paper threads.

Note 9: Issues with statistical strength

As was discussed in Subheading 1.3, the data determined from Pyrosequencing® is accurate enough to allow discrimination of subtle biological effects, involving small changes to imprinting levels in response to various stimuli, to differences between tissues, or to detect partial imprinting. In plants, this may have particular power for identifying PEGs. As described above, using linear regression analysis of PCR performed on gDNA mixtures containing predetermined quantities of each parental allele to generate a standard curve may be a powerful approach for the accurate comparison of the degree of imprinting of partially imprinted alleles [48]. If the degree of imprinting is to be used as a marker for phenotypic trait, then particular care in subsequent association analysis is required. For a discussion of this and the treatment of multiple testing issues, we refer to reviews from the animal literature [49]. A good example of the statistical power possible for well-designed studies of uniparental gene expression is the study of Wang and colleagues [50] in which a "small but consistent and highly statistically significant excess tendency to under-express the paternal X chromosome" was discovered in cDNA of mouse brain.

## 5 Extensions of the Pyrosequencing® Protocol

### 5.1 Use of Pyrosequencing® to Determine Extent of Methylation

Pyrosequencing® has a further application to the study of imprinting which allows insights into the mechanism of uniparental expression, by performing QUASEP on bisulfite-treated DNA to quantify the degree of methylation at cytosine residues within a region. This technique is suitable to imprinted loci in which DMRs have been shown to correlate, or to control, uniparental expression. This technique has been widely used in the analysis of animal imprinted genes, but not, so far to quantify differences between DMRs in plants. This could be of particular use for validating DMRs predicted by genome-wise scans. It could also be used to quantify changes during development, which are associated with a loss of imprinting, or to determine the extent of natural variation in DMRs, which appears to be present between different *FWA* alleles in *A. thaliana* [51]. Most limitations to the use of Pyrosequencing® for methylation-sensitive QUASEP are likely to be associated with the use of bisulfite-treated DNA, which may introduce various problems. A particular issue is the need to design primers suitable for amplifying products from repetitive methylated regions. Some of the methods for improving the success of Pyrosequencing® in such instances have been comprehensively dealt with previously [52].

By applying Pyrosequencing® to bisulfite-treated DNA, it has been possible to determine the degree of methylation in DMRs controlling the uniparental expression of human imprinted genes such as IGF2. For example, it could be shown that ingestion of folic acid during before or during pregnancy was associated with hypomethylation of two DMRs, significant at $p=0.03$ and $p=0.04$, respectively [53]. In addition to the study of different tissues, human imprinted genes have also been studied at multiple developmental stages in order to demonstrate the stability of imprinting through growth and differentiation, or in different states of disease [54]. As with all such studies, the use of suitably sensitive assays and controls such as those suggested by Tost and Gut was used to provide sufficient statistical power to distinguish subtle effects [52]. The use of 50 such assays, which had previously been subject to rigorous validation procedures, was necessary to allow Woodfine and colleagues to determine that the levels of somatic variation exhibited maternally expressed imprinted genes was higher than that exhibited by paternally expressed ones, for example [54].

Such applications will be amenable to dissecting the impact of epigenetic mutations, environmental effects, hybridization, or other challenges to the control of imprinted plant genes in cases where DMRs associated with imprinting are well known. This could particularly be the case for *FWA* and other members of the *HDG* family of HD-ZIP IV TFs [7]. However, it should also be added that of >50 other endosperm expressed genes associated with DMRs, only a further two could be determined to be significantly uniparentally expressed by RNA-seq [11]. Hence, the associations between DMRs and imprinted loci are less well established in angiosperms than in mammals. A further issue is that where DMRs have been identified, they may be widely dispersed or far distal to the coding sequence, to the extent that known DMRs likely to act as ICRs for *MEA* could not be identified by bisulfite sequencing as they were too far upstream [7]. DMRs, which have only been located within a large genomic area, are intrinsically less suitable for analysis by Pyrosequencing®, as many biotinylated PCR products would need to be generated to cover the possible target area. For these regions, we consider that Pyrosequencing® is only likely to prove valuable for the de novo identification of DMRs if good a priori candidate regions can be demonstrated. We identified five predicted DMRs associated with MEGs expressed from either the endosperm or seed coat [13], for example.

Analysis of DNA methylation associated with imprinted loci may be particularly amenable to study by QUASEP as, again, it can exist at different levels, from complete hypo- to complete hypermethylation, and anything in between. Even for genes which are endogenously entirely imprinted in one direction or the other, certain treatments may cause partial reductions in imprinting. For DNA methylation associated with various imprinted genes involved in mouse reproduction, including H19, partial loss of endogenous patterns was observed following treatment with an endocrine-disrupting chemical, methoxychlor [55]. Similar effects can be both observed and quantified during cell culturing: in vitro fertilized mouse embryos have defective DMRs associated with the H19-Igf2 locus when analyzed at the blastocyst stage, but this was rescued after in vitro culture—this led to the development of phenotypically normal embryos with altered expression levels of imprinted genes [56]. The possibility that imprinting levels or the epigenetic marks associated with it could be adjusted, either chemically or via cell culture techniques, has yet to be explored in plants. Given that some reports have suggested that plant imprinted genes are enriched in plant growth regulator response (PGRs) GO terms [14, 15], it is possible that PGRs which act during seed development could have such effects.

As before, it is possible to combine analysis of differential methylation with comparisons between different tissues. Again, these studies may be hindered by the need to generate suitable material. For example, it has only recently become possible to apply studies of differential methylation to preimplantation embryos such as blastocysts [57]. This study also indicated differences between those genes marked with differential methylation in the blastocyst and in embryonic stem cells, which are more easily obtained. In plants, the clear parallel would be the challenging matter of generating sufficiently pure endosperm extracts.

### 5.2 Analysis of the Effects of Epigenetic Modifiers of Imprinting

QUASEP has been used in the animal literature to compare and contrast the response of different imprinted genes to epigenetic perturbation [58]. Such effects are known for various epigenetic mutants in plants, but quantification has only been performed by semiquantitative approaches (end-point PCR and restriction digestion). If such perturbations affect imprinting status without entirely abolishing it (such as the effects of small RNA-biogenesis pathways described by [59]), such genes effectively resemble induced case of partial imprinting. As such, the effects of these epigenetic mutants can be interrogated in a highly quantitative manner by QUASEP as described for partial imprinting, paying particular attention to ensure sufficient replicates and accuracy to allow meaningful comparisons.

### 5.3 Extending the Libraries of SNPs Available for Imprinting Studies

This chapter has principally been concerned with the characterization of the expression of SNPs, which have already been annotated, as is the case for *A. thaliana*. Such SNPs can be derived from sources curated by TAIR, most notably the PERL collection, Col X L*er* SNP data, SNP Viewer, and Ossowski SNPs. These are particularly suitable for the analysis of crosses involving the reference accession, Col-0, together with one or more of those for which SNPs have been well studied. Hence, recent studies have concentrated on Col-0 X L*er*, or Col-0 X Bur-0. Recent papers describing SNPs in detail as part of the 1,001 Genomes project ought to allow these studies to be extended to other accessions. This should allow the evolutionary distribution of genomic imprinting to be more fully assessed. A particular point is that most studies to date have involved crosses to Col-0, and it is possible that analysis of hybrid endosperm, which does not involve this accession, might identify rather different panels of imprinted genes. Pyrosequencing® is also, however, useful for the identification of SNPs, which have not hitherto been identified.

Procedures for identifying previously undescribed SNPs, including computational pipelines for handling of derived data, have recently been described with a focus on two accessions of *Z. mays* [60] and will not be mentioned in detail here. Such pipelines may provide easy tools for moving from partially sequenced genomes, or even EST resources, to identification of SNPs, which may facilitate identification of imprinted genes in other taxa. At present, our lack of data concerning the extent of imprinting in most plant groups, including any dicots apart from *A. thaliana* and its closest relatives, is a significant hindrance to the understanding imprinting from an evolutionary perspective. A further elaboration of such techniques has recently been suggested by Lin and colleagues who have developed an algorithm for identification of de novo SNPs from pooled gDNA samples, in comparison to a known reference [61]. One final consideration is that protocols have been proposed for performing QUASEP on many genes using a three-primer protocol, which avoids the need to generate a separate biotinylated primer for each assay [62]. If it is intended to screen many target genes for predicted monoallelic or imprinted expression, such adaptations may be required to ensure that the project remains competitively priced. If this approach is taken, particular optimization of the preliminary PCR may well be necessary and is discussed in detail by the authors [62].

## 6 Conclusions

The identification of de novo SNPs and the analysis of DMRs, which, in plants especially, may be very large, will always be limited by the relatively short read-lengths currently achieved by QUASEP, which has been identified as a major general limitation of the chemistry [63]. Nevertheless, any application which requires analysis of SNPs—or carefully defined DMRs—associated with candidate imprinted genes in different tissues and genetic backgrounds—may be enriched by the addition of quantitative data. If used with sufficient replication, the use of Pyrosequencing® in this way will increasingly make it possible to distinguish between cases of partial imprinting, to identify partial modifiers of imprinted expression, and to compare the relative levels of uniparental bias exhibited by different tissues.

## References

1. Garnier O, Laoueille-Duprat SL, Spillane C (2008) Genomic imprinting in plants. Epigenetics 3:14–20
2. Barlow DP (2011) Genomic imprinting: a mammalian epigenetic discovery model. Annu Rev Genet 45:379–403
3. Köhler C, Wolff P, Spillane C (2012) Epigenetic mechanisms underlying genomic imprinting in plants. Annu Rev Plant Biol 63:331–352
4. Villar CBR, Erilova A, Makarevich G, Trosch R, Köhler C (2009) Control of *PHERES1* imprinting in Arabidopsis by direct tandem repeats. Mol Plant 2:654–660
5. Luo M, Platten D, Chaudhury A, Peacock WJ, Dennis ES (2009) Expression, imprinting, and evolution of rice homologs of the Polycomb Group genes. Mol Plant 2:711–723
6. Day RC, Herridge RP, Ambrose BA, Macknight RC (2008) Transcriptome analysis of proliferating Arabidopsis endosperm reveals biological implications for the control of syncytial division, cytokinin signaling, and gene expression regulation. Plant Physiol 148:1964–1984
7. Gehring M, Bubb KL, Henikoff S (2009) Extensive demethylation of repetitive elements during seed development underlies gene imprinting. Science 324:1447–1451
8. Hsieh T-F, Ibarra CA, Silva P, Zemach A, Eshed-Williams L, Fischer RL, Zilberman D (2009) Genome-wide demethylation of Arabidopsis endosperm. Science 324:1451–1454
9. Haun WJ, Laoueillé-Duprat SL, O'Connell MJ, Spillane C, Grossniklaus U, Phillips AR, Kaeppler SM, Springer NM (2007) Genomic imprinting, methylation and molecular evolution of maize Enhancer of zeste (Mez) homologs. Plant J 49:325–337
10. Spillane C, Schmid KJ, Laoueille-Duprat SL, Pien S, Escobar-Restrepo J-M, Baroux C, Gagliardini V, Page DR, Wolfe KH, Grossniklaus U (2007) Positive darwinian selection at the imprinted *MEDEA* locus in plants. Nature 450:349–352
11. Gehring M, Missirian V, Henikoff S (2011) Genomic analysis of parent-of-origin allelic expression in *Arabidopsis thaliana* seeds. PLoS One 6:e23687
12. Hsieh T-F, Shin J, Uzawa R, Silva P, Cohen S, Bauer MJ, Hashimoto M, Kirkbride RC, Harada JJ, Zilberman D, Fischer RL (2011) Regulation of imprinted gene expression in Arabidopsis endosperm. Proc Natl Acad Sci U S A 108:1755–1762
13. McKeown PC, Laouielle-Duprat SL, Prins P, Wolff P, Schmid M, Donoghue MTA, Fort A, Duszynska D, Comte A, Lao NT, Wennblom T, Smant G, Kohler C, Grossniklaus U, Spillane C (2011) Identification of imprinted genes subject to parent-of-origin specific expression in *Arabidopsis thaliana* seeds. BMC Plant Biol 11:113
14. Wolff P, Weinhofer I, Seguin J, Roszak P, Beisel C, Donoghue MTA, Spillane C, Nordborg M, Rehmsmeier M, Köhler C (2011) High-resolution analysis of parent-of-origin allelic expression in the Arabidopsis endosperm. PLoS Genet 7:e1002126
15. Luo M, Taylor JM, Spriggs A, Zhang H, Wu X, Russell S, Singh M, Koltunow A (2011) A genome-wide survey of imprinted genes in rice seeds reveals imprinting primarily occurs in the endosperm. PLoS Genet 7:e1002125
16. Waters AJ, Makarevitch I, Eichten SR, Swanson-Wagner RA, Yeh C-T, Xu W, Schnable PS, Vaughn MW, Gehring M, Springer NM (2011) Parent-of-Origin effects on gene expression and DNA methylation in the maize endosperm. Plant Cell 23:4221–4233
17. Zhang M, Zhao H, Xie S, Chen J, Xu Y, Wang K, Zhao H, Guan H, Hu X, Jiao Y, Song W, Lai J (2011) Extensive, clustered parental imprinting of protein-coding and noncoding RNAs in developing maize endosperm. Proc Natl Acad Sci U S A 108:20042–20047
18. O'Connell M, Loughran N, Walsh T, Donoghue MTA, Schmid K, Spillane C (2010) A phylogenetic approach to test for evidence of parental conflict or gene duplications associated with protein-encoding imprinted orthologous genes in placental mammals. Mamm Genome 21:486–498
19. Haig D, Westoby M (1991) Genomic imprinting in endosperm—its effect on seed development in crosses between species, and between different ploidies of the same species, and its implications for the evolution of apomixis. Philos Trans R Soc Lond B Biol Sci 333:1–13
20. Moore T, Mills W (2008) Evolutionary theories of imprinting— enough already! In: Wilkins JF (ed) Genomic imprinting, vol 626, Advances in experimental medicine and biology. Springer, New York, pp 116–122
21. Brandvain Y, Haig D (2005) Divergent mating systems and parental conflict as a barrier to hybridization in flowering plants. Am Nat 166:330–338
22. Wolf JB (2009) Cytonuclear interactions can favor the evolution of genomic imprinting. Evolution 63:1364–1371
23. Köhler C, Weinhofer-Molisch I (2010) Mechanisms and evolution of genomic imprinting in plants. Heredity 105:57–63
24. Marsh S (2007) Pyrosequencing® applications. In: Marsh S (ed) Pyrosequencing® protocols,

vol 373, Methods in molecular biology. Humana, New York, pp 15–23
25. Ronaghi M, Uhlen M, Nyren P (1998) A sequencing method based on real-time pyrophosphate. Science 281:363
26. Ronaghi M (2001) Pyrosequencing sheds light on DNA sequencing. Genome Res 11:3–11
27. Novais RC, Thorstenson YR (2011) The evolution of Pyrosequencing® for microbiology: from genes to genomes. J Microbiol Methods 86:1–7
28. Wang H, Elbein SC (2007) Detection of allelic imbalance in gene expression using Pyrosequencing®. In: Marsh S (ed) Methods in molecular biology, vol 373. Humana, Totowa, pp 157–175
29. Proudhon C, Bourc'his D (2010) Identification and resolution of artifacts in the interpretation of imprinted gene expression. Brief Funct Genomics 9:374–384
30. Raissig MT, Baroux C, Grossniklaus U (2011) Regulation and flexibility of genomic imprinting during seed development. Plant Cell 23:16–26
31. Peters J, Williamson CM (2008) Control of imprinting at the *Gnas* cluster. In: Wilkins JF (ed) Genomic imprinting, vol 626, Advances in experimental medicine and biology. Springer, New York, pp 16–26
32. Klenke S, Siffert W, Frey UH (2011) A novel aspect of GNAS imprinting: Higher maternal expression of Gαs in human lymphoblasts, peripheral blood mononuclear cells, mammary adipose tissue, and heart. Mol Cell Endocrinol 341:63–70
33. Schmid M, Davison TS, Henz SR, Pape UJ, Demar M, Vingron M, Scholkopf B, Weigel D, Lohmann JU (2005) A gene expression map of *Arabidopsis thaliana* development. Nat Genet 37:501–506
34. Zimmermann P, Hirsch-Hoffmann M, Hennig L, Gruissem W (2004) GENEVESTIGATOR. Arabidopsis microarray database and analysis toolbox. Plant Physiol 136:2621–2632
35. Le BH, Cheng C, Bui AQ, Wagmaister JA, Henry KF, Pelletier J, Kwong L, Belmonte M, Kirkbride R, Horvath S, Drews GN, Fischer RL, Okamuro JK, Harada JJ, Goldberg RB (2010) Global analysis of gene activity during Arabidopsis seed development and identification of seed-specific transcription factors. Proc Natl Acad Sci U S A 107:8063–8070
36. Berkowicz EW, Magee DA, Sikora KM, Berry DP, Howard DJ, Mullen MP, Evans RD, Spillane C, MacHugh DE (2011) Single nucleotide polymorphisms at the imprinted bovine insulin-like growth factor 2 ( IGF2) locus are associated with dairy performance in Irish Holstein-Friesian cattle. J Dairy Res 78:1–8
37. Magee DA, Berry DP, Berkowicz EW, Sikora KM, Howard DJ, Mullen MP, Evans RD, Spillane C, MacHugh DE (2011) Single nucleotide polymorphisms within the bovine DLK1-DIO3 imprinted domain are associated with economically important production traits in cattle. J Hered 102:94–101
38. Sikora KM, Magee DA, Berkowicz EW, Berry DP, Howard DJ, Mullen MP, Evans RD, MacHugh DE, Spillane C (2011) DNA sequence polymorphisms within the bovine guanine nucleotide-binding protein Gs subunit alpha (Gsalpha)-encoding (GNAS) genomic imprinting domain are associated with performance traits. BMC Genet 12:4
39. Fakhrai-Rad H, Pourmand N, Ronaghi M (2002) Pyrosequencing™: An accurate detection platform for single nucleotide polymorphisms. Hum Mutat 19:479–485
40. Wang X, Sun Q, McGrath SD, Mardis ER, Soloway PD, Clark AG (2008) Transcriptome-wide identification of novel imprinted genes in neonatal mouse brain. PLoS One 3:e3839
41. Michaels SD, Amasino RM (1998) A robust method for detecting single-nucleotide changes as polymorphic markers by PCR. Plant J 14:381–385
42. Seoighe C, Nembaware V, Scheffler K (2006) Maximum likelihood inference of imprinting and allele-specific expression from EST data. Bioinformatics 22:3032–3039
43. Ruf N, Bähring S, Galetzka D, Pliushch G, Luft FC, Nürnberg P, Haaf T, Kelsey G, Zechner U (2007) Sequence-based bioinformatic prediction and QUASEP identify genomic imprinting of the KCNK9 potassium channel gene in mouse and human. Hum Mol Genet 16:2591–2599
44. Bischoff SR, Tsai S, Hardison N, Motsinger-Reif AA, Freking BA, Nonneman D, Rohrer G, Piedrahita JA (2009) Characterization of conserved and nonconserved imprinted genes in swine. Biol Reprod 81:906–920
45. Shirzadi R, Andersen ED, Bjerkan KN, Gloeckle BM, Heese M, Ungru A, Winge P, Koncz C, Aalen RB, Schnittger A, Grini PE (2011) Genome-wide transcript profiling of endosperm without paternal contribution Identifies parent-of-origin–dependent regulation of *AGAMOUS-LIKE36*. PLoS Genet 7:e1001303
46. Gharizadeh B, Akhras M, Nourizad N, Ghaderi M, Yasuda K, Nyrén P, Pourmand N (2006) Methodological improvements of pyrosequencing technology. J Biotechnol 124:504–511
47. Mosher RA, Melnyk CW, Kelly KA, Dunn RM, Studholme DJ, Baulcombe DC (2009) Uniparental expression of PolIV-dependent

siRNAs in developing endosperm of Arabidopsis. Nature 460:283–286
48. Sun A, Ge J, Siffert W, Frey UH (2004) Quantification of allele-specific G-protein [beta]3 subunit mRNA transcripts in different human cells and tissues by Pyrosequencing. Eur J Hum Genet 13:361–369
49. Magee DA, Berkowicz EW, Sikora KM, Berry DP, Park SDE, Kelly AK, Sweeney T, Kenny DA, Evans RD, Wickham BW, Spillane C, MacHugh DE (2010) A catalogue of validated single nucleotide polymorphisms in bovine orthologs of mammalian imprinted genes and associations with beef production traits. Animal 4:1958–1970
50. Wang X, Soloway PD, Clark AG (2010) Paternally biased X inactivation in mouse neonatal brain. Genome Biol 11:R79
51. Fujimoto R, Kinoshita Y, Kawabe A, Kinoshita T, Takashima K, Nordborg M, Nasrallah ME, Shimizu KK, Kudoh H, Kakutani T (2008) Evolution and control of imprinted *FWA* genes in the genus Arabidopsis. PLoS Genet 4: e1000048
52. Tost J, Gut IG (2007) DNA methylation analysis by pyrosequencing. Nat Protoc 2: 2265–2275
53. Hoyo C, Murtha AP, Schildkraut JM, Jirtle RL, Demark-Wahnefried W, Forman MR, Iversen ES, Kurtzberg J, Overcash F, Huang Z, Murphy SK (2011) Methylation variation at *IGF2* differentially methylated regions and maternal folic acid use before and during pregnancy. Epigenetics 6:928–936
54. Woodfine K, Huddleston JE, Murrell A (2011) Quantitative analysis of DNA methylation at all human imprinted regions reveals preservation of epigenetic stability in adult somatic tissue. Epigenetics Chromatin 4:1–13
55. Stouder C, Paoloni-Giacobino A (2011) Specific transgenerational imprinting effects of the endocrine disruptor methoxychlor on male gametes. Reproduction 141:207–216
56. Fauque P, Ripoche M-A, Tost J, Journot L, Gabory A, Busato F, Le Digarcher A, Mondon F, Gut I, Jouannet P, Vaiman D, Dandolo L, Jammes H (2010) Modulation of imprinted gene network in placenta results in normal development of in vitro manipulated mouse embryos. Hum Mol Genet 19:1779–1790
57. Huntriss J, Woodfine K, Huddleston JE, Murrell A, Rutherford AJ, Elder K, Khan AA, Hemmings K, Picton H (2011) Quantitative analysis of DNA methylation of imprinted genes in single human blastocysts by pyrosequencing. Fertil Steril 95:2564–2567
58. Weaver JR, Sarkisian G, Krapp C, Mager J, Mann MRW, Bartolomei MS (2010) Domain-specific response of imprinted genes to reduced *DNMT1*. Mol Cell Biol 30:3916–3928
59. Bratzel F, Yang C, Angelova A, López-Torrejón G, Koch M, del Pozo JC, Calonje M (2012) Regulation of the new Arabidopsis imprinted gene *AtBMI1C* requires the interplay of different epigenetic mechanisms. Mol Plant 5:260–269
60. Barbazuk WB, Schnable PS (2011) SNP discovery by transcriptome pyrosequencing. In: Lu C, Browse J, Wallis JG (eds) Methods in molecular biology, vol 729. Humana, Totowa, pp 225–246
61. Lin Y-S, Liu F-GR, Wang T-Y, Pan C-T, Chang W-T, Li W-H (2011) A simple method using Pyrosequencing™ to identify de novo SNPs in pooled DNA samples. Nucleic Acids Res 39:e28
62. Royo JL, Hidalgo M, Ruiz A (2007) Pyrosequencing protocol using a universal biotinylated primer for mutation detection and SNP genotyping. Nat Protoc 2:1734–1739
63. Ahmadian A, Ehn M, Hober S (2006) Pyrosequencing: history, biochemistry and future. Clin Chim Acta 363:83–94

# Chapter 7

# Endosperm-Specific Chromatin Profiling by Fluorescence-Activated Nuclei Sorting and Chip-on-Chip

Isabelle Weinhofer and Claudia Köhler

## Abstract

Cell-type-specific analysis of gene expression and chromatin profiling requires the isolation of discrete cell populations from complex pools. However, until now this most critical step has been labor intensive and technical challenging. Here, we describe a rapid protocol based on fluorescence-activated cell sorting (FACS) for cell-type-specific RNA and chromatin profiling. We detail how to isolate nuclei from Arabidopsis inflorescence and silique homogenates and how to purify endosperm nuclei labeled by nuclear-targeted green fluorescent protein using FACS. The purified fluorescent endosperm nuclei can be further used for chromatin immunoprecipitation (ChIP) followed by hybridization to high-resolution whole-genome tiling microarrays (ChIP-on-chip) or transcriptional profiling.

**Key words** Fluorescence-activated nuclei sorting, Endosperm, Chromatin profiling, ChIP-on-chip

## 1 Introduction

In flowering plants like Arabidopsis, a unique double fertilization event gives rise to a diploid zygote and a triploid endosperm. The endosperm is a tissue with unequal parental genomic contribution, which, like the placenta in mammals, serves as a food source for the developing embryo and is essential for normal embryo development [1]. Endosperm development is regulated by Polycomb group (PcG) proteins that establish trimethylation marks on lysine 27 of histone H3 (H3K27me3; for a protocol to determine the extent of this by chromatin immunoprecipitation (ChIP), see the chapter of Song, Rutjens and Dean elsewhere in this volume). Loss of PcG function in the endosperm causes endosperm overproliferation and cellularization failure, eventually leading to seed abortion [2]. The endosperm is an inaccessible tissue as it is surrounded by the maternally derived seed coat. To understand the special developmental trajectory of the endosperm and the functional role of PcG proteins in this process, we developed a new isolation procedure of endosperm nuclei that is based on fluorescence-activated

Charles Spillane and Peter C. McKeown (eds.), *Plant Epigenetics and Epigenomics: Methods and Protocols*, Methods in Molecular Biology, vol. 1112, DOI 10.1007/978-1-62703-773-0_7, 

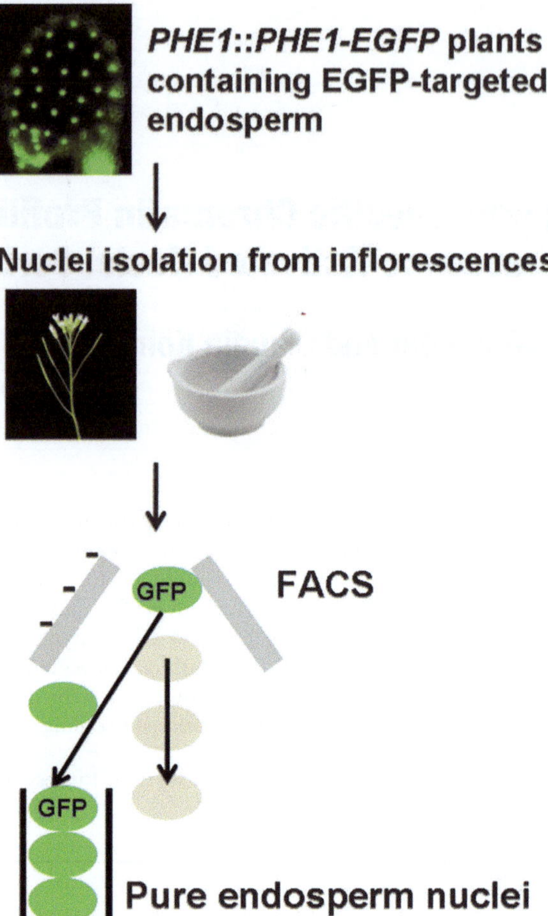


**Fig. 1** Flow chart describing the isolation of purified endosperm nuclei. Nuclei are isolated from siliques of PHE1::*PHE1-EGFP* plants expressing EGFP restricted to endosperm nuclei. Using Fluorescent-Activated Cell Sorting (FACS), EGFP-positive endosperm nuclei can be specifically isolated

cell sorting (FACS; Fig. 1; as used in ref. 3). Although previous studies reported successful manual dissection of the endosperm [4, 5], this approach is highly labor intensive and not applicable to seeds of earlier developmental stages [1–4 days after pollination (DAP)]. In our protocol, we used a transgenic Arabidopsis line that expresses EGFP as a nuclear-targeted translational fusion of *PHERES1* (*PHE1*) [6] under the transcriptional control of the *PHE1* promoter (PHE1::*PHE1-EGFP*) und 3 kb regulatory 3′ sequences. *PHE1* is a gene that is specifically expressed in endosperm nuclei from 1 DAP to 4 DAP [3]. We modified the protocol described in ref. 7 to isolate nuclei from inflorescence homogenates and purified EGFP-labeled endosperm nuclei by FACS. We further detail how to isolate RNA and chromatin from these nuclei for gene expression analysis and ChIP, respectively, and hybridization of ChIP-DNA to high-resolution whole-genome tiling microarrays (ChIP-on-chip).

## 2 Materials

### 2.1 Nuclei Isolation

1. Plant material: Tissue from inflorescences and siliques of PHE1::*PHE1-EGFP* plants, *35S::H3.2-YFP* plants (positive control), wild-type plants (negative control).
2. *Honda buffer*: 2.5 % (w/v) Ficoll 400 (add slowly while stirring); 5 % (w/v) dextran T40; 0.4 M sucrose; 25 mM Tris–HCl (pH 7.4); 10 mM $MgCl_2$; sterile filtrate and store at 4 °C; before use add 10 mM ß-mercaptoethanol, 100 μg/mL PMSF, and 1 μg/mL Pepstatin A.
3. 20 % (v/v) Triton X-100 in water.
4. Phosphate-buffered saline (PBS).
5. *Protease inhibitor solution* (*25×*): Dissolve one tablet complete (Roche) per 2 mL sterile distilled water. Can be stored at −20 °C.
6. 100-μm pore size nylon mesh (Lanz-Anliker AG).
7. CellTrics 30 μm filters (Partec).

### 2.2 Cross-linking of Isolated Nuclei

1. Solution of 1× PBS and formaldehyde (1 %). Formaldehyde is highly toxic and volatile, avoid inhalation and skin contact.
2. 1.25 M Glycine.
3. 1× PBS containing 1× Protease inhibitor solution.

### 2.3 Fluorescence-Activated Sorting of Nuclei

1. Polystyrene 5 mL round bottom tubes (BD Falcon).
2. 6 μm beads (Becton Dickinson).
3. FACS Aria II (Becton Dickinson).
4. Propidium iodide (PI, Sigma-Aldrich). PI is a known mutagen, avoid skin contact.
5. *2× Nuclei Lysis buffer*: 100 mM Tris–HCl (pH 8.1), 20 mM EDTA, 2 % (w/v) SDS, before use add 1× Protease inhibitor solution.
6. RLT-lysis buffer from RNeasy Mini Kit (Qiagen).

### 2.4 Gene Expression Analysis of Sorted Nuclei

1. RNeasy Mini Kit (Qiagen).
2. RNase-free DNase Set (Qiagen).
3. RevertAid First strand cDNA synthesis kit (Fermentas).
4. Fast-SYBR-mix (Applied Biosystems).
5. 7500 Fast Real-Time PCR system (Applied Biosystems).

#### 2.5 Chromatin Immunoprecipitation (ChIP) (Adopted from [8])

1. BioruptorTM200 Sonicator (Diagenode).
2. 5 M NaCl.
3. DNase-free RNase (Fermentas).
4. Proteinase K (Roche).
5. QIAquick PCR purification kit (Qiagen).
6. QIAquick MinElute PCR purification kit (Qiagen).
7. NanoDrop (Thermo Scientific).
8. Lyophilized Staph A cells (CalBiochem).
9. *1× Dialysis buffer*: 2 mM EDTA, 50 mM Tris–HCl (pH 8.0), and 0.2 % (w/v) Sarkosyl (omit for monoclonal antibodies).
10. *1× Dialysis buffer* without sarkosyl.
11. 1× PBS containing 3 % (w/v) SDS and 10 % (w/v) β-mercaptoethanol.
12. Yeast tRNA (Promega).
13. BSA.
14. 100 mM PMSF in ethanol.
15. 0.6 mL low retention siliconized tubes (Fisher).
16. Anti-H3K27me3 (Lucerna Chem AG).
17. IgG (Santa Cruz Biotechnology).
18. *IP dilution buffer*: 0.01 % (w/v) SDS, 1.2 mM EDTA, 16.7 mM Tris–HCl (pH 8.1), and 167 mM NaCl; before use, add 1.1 % Triton X-100 and protease inhibitors.
19. *IP wash buffer*: 100 mM Tris–HCl (pH 9.0; for monoclonal antibodies: pH 8.0), 500 mM LiCl, and 1 % deoxycholic acid (Fisher); before use add 1 % Igepal (v/v; Sigma-Aldrich).
20. *Elution buffer*: 50 mM $NaHCO_3$ and 1 % SDS.

#### 2.6 Amplification of Immunoprecipitated DNA

1. WGA4 GenomePlex Single Cell Whole genome Amplification Kit (Sigma-Aldrich).
2. dUTP (Roche).

#### 2.7 Labeling of Immunoprecipitated DNA

1. GeneChip WT Terminal Labeling kit (Affymetrix).
2. RNA Nano 1000 kit (Agilent).
3. 2100 Bioanalyzer lab-on-chip platform (Agilent).
4. AGRONOMICS1 arrays [9].

## 3 Methods

#### 3.1 Nuclei Isolation

All steps should be performed on ice, precool buffer and plastic tubes on ice.

1. Freeze 3.5 g inflorescence tissue containing siliques with liquid nitrogen and homogenize the tissue by grinding with mortar and pestle. Transfer homogenate to a 50-mL plastic tube and add 4-mL Hondabuffer.
2. Filter homogenate through a 100-μm nylon mesh into a new 50-mL plastic tube, wash nylon mesh with additional 3-mL Hondabuffer (*see* **Note 1**).
3. Filter homogenate through 30-μm CellTrics filter into a 15-mL plastic tube.
4. Add Triton X-100 to the filtrate to a final concentration of 0.5 %, mix gently and incubate for 15 min on ice.
5. Centrifuge at 1,500×*g* for 5 min, 4 °C. Remove the supernatant and wash the nuclei pellet with 2 mL ice-cold Hondabuffer containing 0.1 % Triton X-100. Centrifuge at 1,500×*g* for 5 min at 4 °C (*see* **Note 2**).

### 3.2 Cross-linking of Isolated Nuclei

All steps should be performed on ice, using precooled buffer and keeping plastic tubes on ice.

1. Remove the supernatant and gently resuspend the pellet in 1 mL 1× PBS containing 1 % formaldehyde. Mix gently and incubate for 8 min on ice. Stop cross-link by addition of glycine to 125 mM final concentration. Incubate for 5 min on ice (*see* **Note 2**).
2. Centrifuge at 1,500×*g* for 5 min at 4 °C. Remove the supernatant and wash the pellet with 1 mL ice-cold 1× PBS containing protease inhibitors. Repeat this step (*see* **Note 2**).
3. Remove the supernatant and resuspend the nuclei pellet in 0.5 mL of ice-cold 1× PBS containing protease inhibitors. Transfer solution to FACS tubes (*see* **Note 2**).

### 3.3 Fluorescence-Activated Sorting of Nuclei

1. Add Propidium Iodide (PI) to the nuclei solution to a final concentration of 1 μg/mL. Mix briefly by vortexing.
2. Perform biparametric flow analysis of GFP fluorescence versus DNA content (*see* **Notes 3** and **4**).

### 3.4 Gene Expression Analysis of Sorted Nuclei

1. Sort a minimum of 100,000 nuclei directly into 450 μL of Qiagen RLT lysis buffer and isolate RNA according to Qiagen RNeasy Mini Kit instructions. For quantitative RT-PCR, include the Qiagen RNase-free DNase Set to remove residual DNA and reverse transcribe the RNA using the Fermentas First strand cDNA synthesis kit according to the manufacturer's recommendation. Use gene-specific primers and the Fast-SYBR-mix from Applied Biosystems to perform quantitative PCR on a Real-Time PCR system.

### 3.5 Chromatin Immunoprecipitation

1. Sort approximately 100,000 nuclei directly into 2× ChIP lysis buffer containing protease inhibitors. Transfer nuclei solution to a 1.5-mL plastic tube, incubate 10 min on ice and flash freeze in liquid nitrogen. Sorted nuclei can be stored at −80 °C until needed (*see* **Notes 4** and **5**).
2. Thaw tubes at room temperature and sonicate the chromatin on ice at high power for 30 s ON, 1 min OFF for a total time of 10 min. To check the correct fragment size of 300 bp, run 2 μg of chromatin on a 1.5 % agarose gel (*see* **Note 6**).
3. Centrifuge at 16,000 × *g* for 10 min at 4 °C. Collect the supernatant and transfer to a fresh 1.5-mL tube. Take 10 μL of chromatin for quantification and flash freeze the rest in liquid nitrogen. Store at −80 °C. *The protocol may be paused at this point.*
4. Chromatin quantification: to 10 μL chromatin, add 90 μL of $dH_2O$ and 10 μL of 5 M NaCl. Boil for 15 min, then add 1 μL of DNase-free RNase. Incubate for 15 min at 37 °C, then add 1 μL of proteinase K and incubate at 67 °C for 15 min. Purify the chromatin using QIAquick columns according to the manufacturer's instruction and measure the DNA concentration by using a NanoDrop or similar.
5. Prepare Staph A cells: resuspend 1 g of lyophilized Staph A cells in 10 mL of 1× dialysis buffer and centrifuge at 9,000 × *g* for 5 min at 4 °C. Resuspend the cell pellet in 10 mL of 1× dialysis buffer and repeat the centrifugation step. Resuspend the pellet in 3 mL of 1× PBS containing 3 % SDS and 10 % β-mercaptoethanol and boil for 30 min. Centrifuge at 9,000 × *g* for 5 min at 4 °C and wash the pellet twice with 1× dialysis buffer. Resuspend the pellet in 4 mL of 1× dialysis buffer and divide into 200 μL aliquots. Snap freeze the aliquots in liquid nitrogen and store them at −80 °C.
6. Pretreat Staph A cells: thaw 1 aliquot of Staph A cells (200 μL) on ice and add 25 μL of yeast tRNA (10 mg/mL) and 25 μL of BSA (10 mg/mL). Incubate the tube on a rotating platform at 4 °C overnight. Centrifuge at 16,000 × *g* for 5 min at 4 °C, remove the supernatant and wash the pellet twice with 1.4 mL of dialysis buffer. Resuspend the cell pellet in 200 μL of dialysis buffer without sarkosyl and add PMSF to a concentration of 1 mM.
7. Preclear the chromatin: add 30 μL of blocked Staph A cells to the chromatin and incubate on a rotating platform for 15 min at 4 °C. Centrifuge at 16,000 × *g* for 10 min at 4 °C. Transfer the supernatant to a new tube and measure the volume.
8. Formation of antibody—chromatin complexes: for the input-DNA control, transfer 10 ng of DNA to a 1.5-mL tube and store at −20 °C. To both the sample and the nonspecific IgG control, transfer 500–700 ng of chromatin to a 0.6 mL

low-retention siliconized tube and adjust volume to 500 μL with IP dilution buffer to which protease inhibitors have been added. Add 1 μg of antibody against H3K27me3 or IgG, respectively, and incubate on a rotating platform at 4 °C overnight.

9. Purification of antibody—chromatin complexes: add 5 μL of blocked Staph A cells to the sample and to the IgG control and incubate on a rotating platform for 15 min. Centrifuge at 14,000 × *g* for 10 min at 4 °C and remove the supernatant. Wash the pellet by resuspending in 250 μL of 1× dialysis buffer, add an additional 250 μL of 1× dialysis buffer and incubate the tubes on a rotating platform for 3 min at 4 °C. Centrifuge at 16,000 × *g* for 3 min at 4 °C and remove the supernatant without aspirating the Staph A cells. Repeat the washing step with 0.5 mL of 1× dialysis buffer and then wash accordingly three times with 0.5 mL of IP wash buffer. Elute the antibody—chromatin complexes by addition of 50 μL of IP elution buffer and vortexing the tubes for 20 min at 4 °C. Centrifuge at 16,000 × *g* for 5 min at room temperature and transfer the supernatant to a fresh tube. Repeat the elution step with additional 50 μL of IP elution buffer and combine both eluates in the same tube. Centrifuge at 16,000 × *g* for 5 min at room temperature and transfer the supernatant to a new tube.
10. Revert the cross-link: add 10 μL of 5 M NaCl (to a final concentration: 0.45 M) and boil the tubes for 15 min. Purify the ChIP-DNA with QIAquick MinElute columns according to the manufacturer's instructions and elute in 10 μL of $ddH_2O$.
11. Prepare input DNA: to the input DNA saved in **step 8**, add $ddH_2O$ to a final volume of 100 μL. Revert cross-link by addition of 10 μL of 5 M NaCl and boiling for 15 min. Add 1 μL of DNase-free RNase and incubate at 37 °C for 15 min, then add 1 μL of Proteinase K and incubate at 67 °C for 15 min. Purify the input DNA using QIAquick MinElute columns according to the manufacturer's instructions and elute in 10 μL of $ddH_2O$. Measure the concentration of the input DNA by using a NanoDrop or similar.

### 3.6 Amplification of Immunoprecipitated DNA

1. Transfer 5 μL of purified ChIP-DNA to a PCR tube (for input DNA, transfer 10 ng), bring volume to 11 μL with $dH_2O$ and amplify ChIP-DNA using the WGA-4 single-cell amplification kit (Sigma-Aldrich, omit **steps 1–5**). For later fragmentation (**step 17**) necessary for microarray hybridization, include 125 μM of dUTP in the PCR amplification reaction. After amplification, purify the DNA with the Qiagen QIAquick PCR purification columns according to the manufacturer's instruction and elute in 50 μL of $ddH_2O$. The typical yield is around 10 μg of amplicon. Amplified ChIP-DNA can be stored at −20 °C until needed.

### 3.7 Labeling of Immunoprecipitated DNA

1. Fragment and label 5 μg of ChIP-DNA using the Affymetrix GeneChip WT Terminal Labeling kit according to the manufacturer's instructions. Confirm the fragmentation using the RNA Nano 1000 kit on a 2100 Bioanalyzer lab-on-chip platform. The average fragment size should be around 90 nucleotides. Hybridize labeled samples (Input, ChIP with anti-H3K27me3, and ChIP with unspecific IgG) from three independent experiments to Affymetrix AGRONOMICS1 arrays [9].

### 3.8 Validation of ChIP-on-Chip Results by Quantitative PCR

1. Dilute amplified ChIP-DNA 1:30 with $ddH_2O$ and use 2 μL per quantitative PCR.

## 4 Notes

1. Do not squeeze the nylon mesh containing the inflorescence homogenate but allow filtration by gravitation only. Squeezing results in transfer of particles that will clog the CellTrics 30-μm filters in **step 3**.
2. To avoid destruction of nuclei, gently resuspend the nuclei pellet with a soft brush. Avoid exaggerated pipetting and/or introduction of air bubbles.
3. Perform biparametric flow analysis of GFP fluorescence versus DNA content on an FACS Aria II or equivalent sorter equipped with a 70-μm flow tip and operated at a sheath pressure of 70 psi. Threshold events on forward scatter sort the sample at an event rate of 15,000 per second. For GFP and PI excitation, use a 488-nm laser and 610/20 nm (PI) or 530/30 (GFP) barrier filters. A typical Arabidopsis nucleus has a radius of around 5 μm, thus defines the position of the nuclei gate using 6 μm beads (Becton Dickinson) and forward (FSC-A) and sidewards scatter (SSC-A). Verify the gate by DAPI-staining of the nuclei and usage of 610/20 nm barrier filters: upon DAPI-staining, a pure nuclei fraction will shift to the right side on the logarithmic scale (Fig. 2a). Establish the position of the sort region by first determining the baseline of green fluorescence using inflorescence nuclei from GFP-negative control plants (e.g., *Ler*, Fig. 2b). Adjust the upper and left- and right-hand boundaries of the sort window by including all nuclei derived from GFP-positive control plants (e.g., 35S::H3.2-YFP, Fig. 2b). Reanalyze sorted GFP-positive nuclei to verify sorting conditions (Fig. 2c).

4. *Critical step*: After sorting, drops containing sorted nuclei tend to be spread across the walls of the FACS tube. Gently rolling the closed FACS tube transfers nuclei drops to the bottom of the tube thus avoiding loss of nuclei by drying.

5. To prevent precipitation of the SDS, do not store the buffer on ice (keep at 4 °C instead).

6. An hour before use, fill the BioruptorTM200 with ice for cooling. Before use, remove the ice with an exception of a thin layer and add ice-cold water until the level advised by the manufacturer's instruction. Use identical 1.5-mL plastic tubes (brand, material, etc.) and fill the tubes with a volume between 100 and 300 μL.

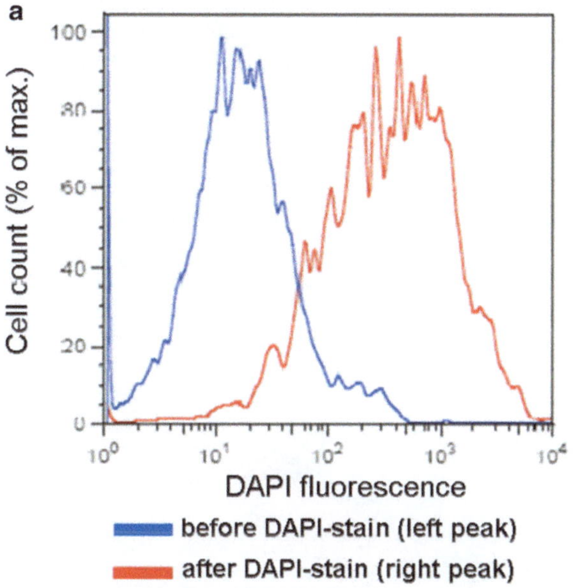


**Fig. 2** Establishment of suitable conditions for GFP sorting. (**a**) The presence of nuclei and purity of the defined nuclei gate was verified by analyzing GFP-positive nuclei isolated from PHE1::*PHE1-EGFP* plants by flow cytometry before (*blue line*, left-hand peak) and after DAPI staining (*red line*, right-hand peak). After addition of DAPI, the whole population of particles present in the defined nuclei gate is shifted to higher DAPI fluorescence, indicating high purity of isolated nuclei. (**b**) Biparametric flow sort analysis of nuclei isolated from wild-type inflorescences (*upper panel*), from 35S::*H3.2-YFP* inflorescences (*middle panel*) and from PHE1::*PHE1-EGFP* inflorescences (*lower panel*). P3 represents the region employed for sorting GFP-negative nuclei. P4 represents the region containing GFP-positive nuclei. (**c**) The purity of isolated GFP-positive nuclei from PHE1::*PHE1-EGFP* plants was verified by reanalysis of the sorted sample. The sorted sample (*green line*, left-hand peak) was clearly enriched for GFP-positive nuclei compared to the unsorted sample (*blue line*, right-hand peak). *Bars* indicate GFP-positive signals. The presence of two peaks is likely contributed to endoreduplication and correspondingly increased GFP signal intensity

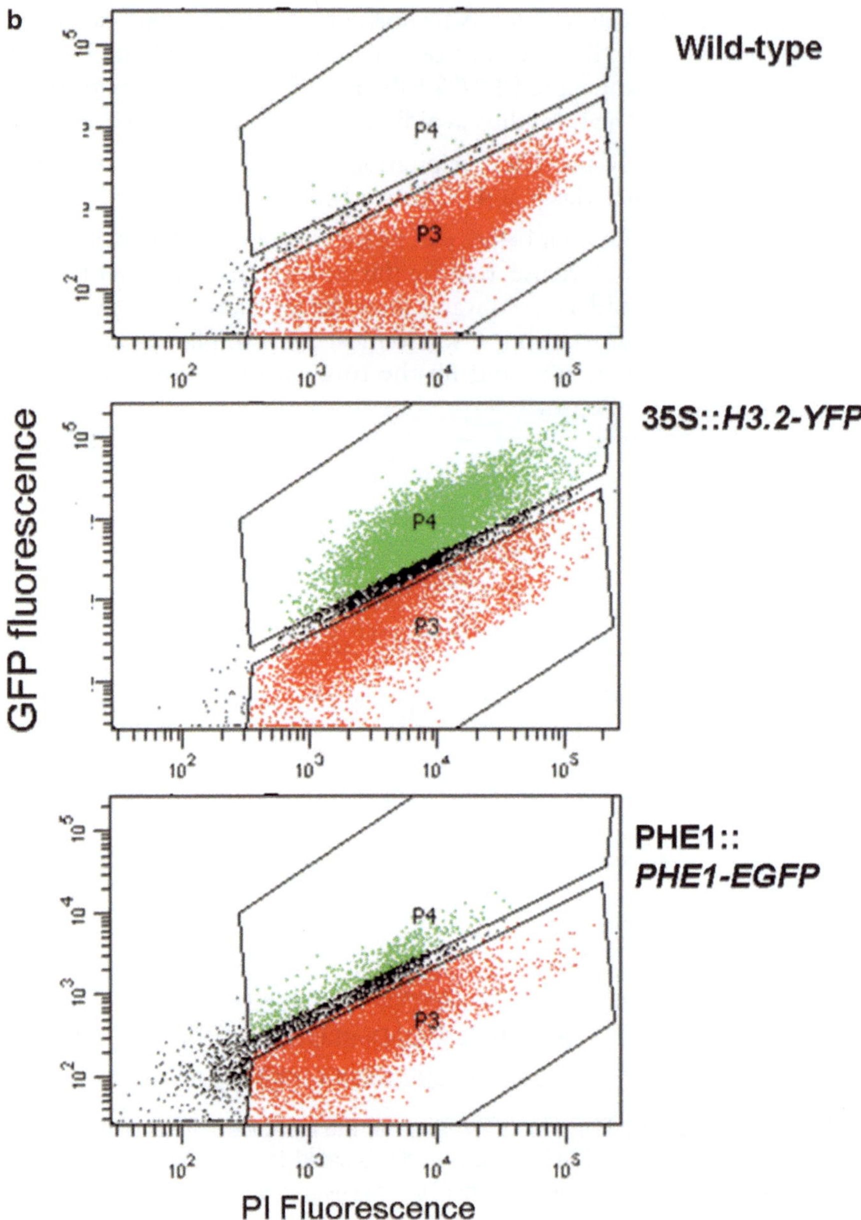


**Fig. 2** (continued)

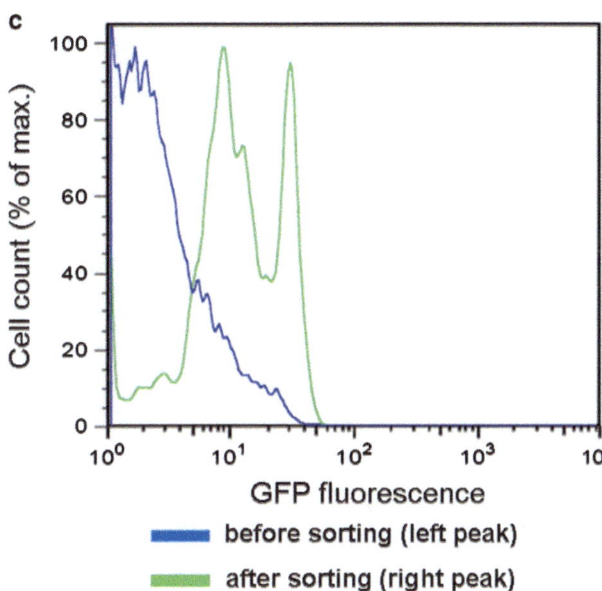


**Fig. 2** (continued)

## References


1. Costa LM, Gutierrez-Marcos JF, Dickinson HG (2004) More than a yolk: the short life and complex times of the plant endosperm. Trends Plant Sci 9:507–514
2. Kohler C, Makarevich G (2006) Epigenetic mechanisms governing seed development in plants. EMBO Rep 7:1223–1227
3. Weinhofer I, Hehenberger E, Roszak P, Hennig L, Kohler C (2010) H3K27me3 profiling of the endosperm implies exclusion of polycomb group protein targeting by DNA methylation. PLoS Genet 6:e1001152
4. Gehring M, Bubb KL, Henikoff S (2009) Extensive demethylation of repetitive elements during seed development underlies gene imprinting. Science 324:1447–1451
5. Kinoshita T, Yadegari R, Harada JJ, Goldberg RB, Fischer RL (1999) Imprinting of the MEDEA polycomb gene in the Arabidopsis endosperm. Plant Cell 11:1945–1952
6. Kohler C, Hennig L, Spillane C, Pien S, Gruissem W, Grossniklaus U (2003) The Polycomb-group protein MEDEA regulates seed development by controlling expression of the MADS-box gene *PHERES1*. Genes Dev 17:1540–1553
7. Weigel D, Glazebrook J (2002) Arabidopsis: a laboratory manual. Cold Spring Harbor Laboratory Press, New York
8. Acevedo LG, Iniguez AL, Holster HL, Zhang X, Green R, Farnham PJ (2007) Genome-scale ChIP-chip analysis using 10,000 human cells. Biotechniques 43:791–797
9. Rehrauer H, Aquino C, Gruissem W, Henz SR, Hilson P, Laubinger S, Naouar N, Patrignani A, Rombauts S, Shu H, Van de Peer Y, Vuylsteke M, Weigel D, Zeller G, Hennig L (2010) AGRONOMICS1: a new resource for Arabidopsis transcriptome profiling. Plant Physiol 152: 487–499

# Chapter 8

# Imaging Sexual Reproduction in Arabidopsis Using Fluorescent Markers


## Mathieu Ingouff



## Abstract

Sexual reproduction in higher plants is a stealth process as most events occur within tissues protected by multiple surrounding cell layers. Female gametes are produced inside the embryo sac surrounded by layers of ovule integument cells. Upon double fertilization, two male gametes are released at one end of the embryo sac and migrate towards their respective female partner to generate the embryo and its feeding tissue, the endosperm, within a seed. Since the early discovery of plant reproduction, advances in microscopy have contributed enormously to our understanding of this process (Faure and Dumas, Plant Physiol 125:102–104, 2001). Recently, live imaging of double fertilization has been possible using a set of fluorescent markers for gametes in Arabidopsis. The following chapter will detail protocols to study male and female gametogenesis and double fertilization in living tissues using fluorescent markers.



**Key words** Arabidopsis, Double fertilization, Fluorescent markers, Gametes, Live imaging


## 1 Introduction

Sexual reproduction in Arabidopsis starts with production of gametes founder cells (microspores and megaspores) through meiosis that enter into a series of mitoses to generate two male gametes in the male gametophyte (the pollen grain) and two females gametes (the egg cell and central cell) and accessory cells in the female gametophyte (the embryo sac) [1–4]. During the double fertilization process, the two male gametes travel through the embryo sac to fuse with their female partner to initiate the development of the embryo and its feeding tissue the endosperm into a seed [5–7]. One of the main technical obstacles to study sexual reproduction is the presence of maternal tissues surrounding the precursor cells of the gametes and the female gametes themselves. Technical hurdles have been recently overcome in Arabidopsis following the recent development of protocols to purify male and female gametes [8, 9] to dynamically image cellular processes within reproductive tissues [10, 11] and the discovery of cell-type-specific fluorescent markers [6, 12].


Charles Spillane and Peter C. McKeown (eds.), *Plant Epigenetics and Epigenomics: Methods and Protocols*,
Methods in Molecular Biology, vol. 1112, DOI 10.1007/978-1-62703-773-0_8, © Springer Science+Business Media New York 2014

## 2 Materials

1. Sterile Murashige and Skoog medium (1/4, pH 5.8) supplemented with 0.1 % gelrite from Duchefa (*see* **Note 1**).
2. Sorbitol from Sigma.
3. Microscope coverslips #0 (22 × 22 × 0.12 mm) from EMS.
4. Standard glass slides.
5. Super fine dissecting tweezers Dumont® N5 from EMS.
6. Fine dissecting tweezers Dumont® N3 from EMS.
7. Hypodermic needles (0.45 × 23 mm) from EMS.
8. Scotch® double-sided tape.
9. Square Petri dishes (120 × 120 × 17 mm) from Dutscher.
10. Nuclear stain: 1 mg/mL DAPI (4,6-diamino-2-phenylindole) in water from Sigma.

## 3 Methods

### 3.1 Sample Preparation to Obtain Microspores, Bicellular and Tricellular Pollen Grains

Male gametogenesis occurs in anthers and starts from a meiosis-derived cell, the microspore that divides to generate the vegetative cell and the generative cell in a bicellular pollen grain (Fig. 1). A second mitosis occurs in the generative cell to form the two male gametes in the tricellular pollen grain. The following protocol is mainly derived from the excellent technical review of McCormick's lab [20].

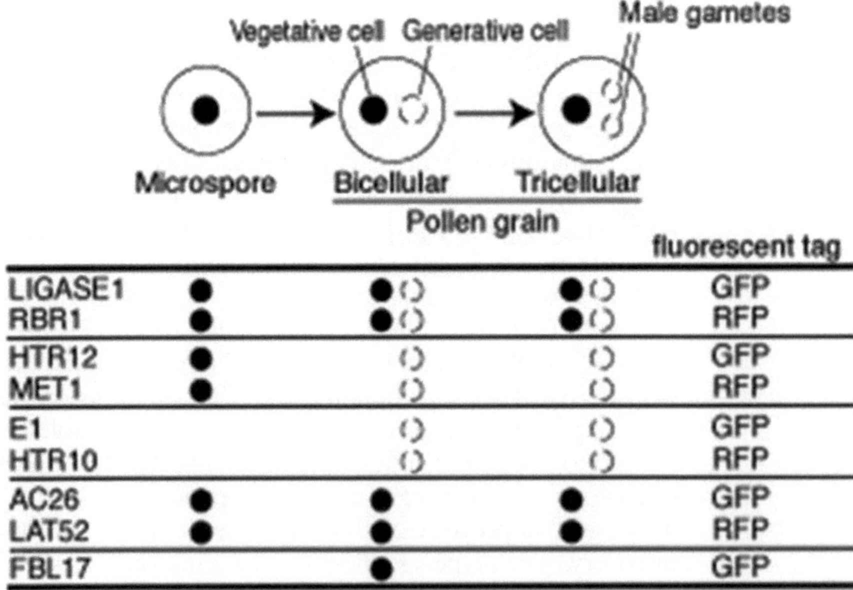


**Fig. 1** Marker lines to study male gametogenesis. This table is not exhaustive but provides a combination of selected fluorescent markers to precisely study male gametogenesis. References corresponding to the mentioned genes and transgenic marker lines: *AC26* [13]; *E1* [14]; *FBL17* [15]; *HTR10* [16]; *HTR12* [16, 17]; *LAT52* [18]; *MET1* [19]. Abbreviations: *GFP* green fluorescent protein, *RFP* red fluorescent protein

1. Collect a flower by the peduncle with a fine forceps No. 3 (DUMONT) (*see* **Note 2**).
2. Affix the selected flower on a ribbon of double-sided adhesive tape placed on a square Petri dish.
3. Under a dissecting microscope fitted with a cold light source, peal off the sepals and petals with a hypodermic needle.
4. Cut off the anthers from the filament.

5a. Place them onto a glass slide containing few drops of sorbitol 80 mM.

5b. Optional: Add few drops of DAPI (4′,6-diamidino-2-phenylindole; 1 mg/mL) and incubate for 5 min at room temperature to stain the chromatin.

6. Cover the sample with a coverslip.
7. Press gently onto the coverslip and start moving circularly the coverslip to release the male gametophytes from the anthers.

### 3.2 Imaging of Microspores, Bicellular and Tricellular Pollen Grains

Standard epifluorescent microscopy is usually sufficient to obtain sharp images for the microspores and developing pollen grains once separated from anthers. Major fluorescent marker lines available for the different stages of male gametogenesis are compiled in Fig. 1. Some of these markers have been used to study cell fate specification during male gametogenesis after introgression in mutant backgrounds [21, 22].

### 3.3 Sample Preparation to Image Different Steps of Megagametogenesis

Female gametogenesis occurs in the embryo sac protected by ovule integuments. Upon meiosis, a surviving megaspore generates an eight-nuclei syncytium after three successive rounds on nuclei divisions. At this stage, a wave of cellularization occurs concomitantly with the migration of two nuclei (the polar nuclei) towards the center of the embryo sac that ultimately fuse. At maturity the seven-celled embryo sac contains four cell types: two synergids, two female gametes (the egg cell and the central cell), and three antipodal cells. As no apparent differentiation occurs before cellularization of the embryo sac, only cell-type-specific fluorescent markers are presented in Fig. 2. Although specific markers of egg cell are scarce, many markers are now available for the other cell types of the embryo sac [23, 24, 30]. These markers can be combined to track abnormal cell differentiation in a mutant background. A comprehensive description of the different steps of megagametogenesis has been presented by Christensen et al. [31].

1. Collect a flower by the peduncle with a fine forceps No. 3 (DUMONT) (*see* **Note 3**).
2. Affix the selected flower on the ribbon of double-sided adhesive tape placed on a square Petri dish.
3. Under a dissecting microscope fitted with a cold light source, slide down a bent hypodermic needle along each side of the pistil to remove the sepals, petals, and stamens.

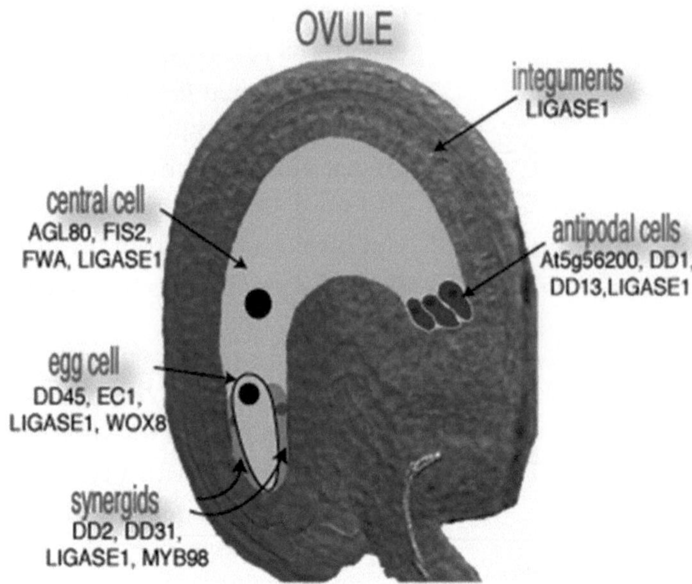


**Fig. 2** Cell-type-specific and general fluorescent markers expressed in the mature female gametophyte. The female gametophyte contains four cell types at maturity: two female gametes (egg cell and central cell), two synergids, and three antipodal cells. Fluorescent markers specific for each of these cell types are presented. References corresponding to the mentioned genes and transgenic marker lines: *AGL80* [18]; At5g56200 [23]; *DD1*, *DD2*, *DD13*, *DD31*, *DD45* [24]; *EC1* [22, 25]; *FIS2* [26]; *FWA* [27]; *LIGASE1* [28]; *MYB98* [29]; *WOX8* [23]

4. Slit open the pistil by passing the bent hypodermic needle just under the replum from top to bottom (*see* **Note 4**).
5. Cut the top of the pistil where stigmatic papillae develop.
6. Incise gently one of the two valves of the pistil placing the hypodermic needle between the valve and the row of ovules.
7. Collect with a superfine forceps No. 5 (DUMONT) the septum with files of ovules attached to it holding firmly the remaining valve with a fine forceps No. 3 (DUMONT).
8. Place the septum onto a glass slide containing few drops of MS medium (1/4, pH 5.8) supplemented with 0.1 % gelrite (*see* **Note 5**).
9. Cover the sample with a coverslip.

### 3.4 Sample Preparation to Image Steps Leading to Double Fertilization

Once a pollen grain lands onto the stigma, its vegetative cell germinates a pollen tube to transport the two male gametes inside the pistil towards an ovule. The two male gametes are released into one of the two synergids and move towards the female gametes. Upon double fertilization, the ovule becomes a seed where the fertilized egg cell generates the zygote and the fertilized central cell

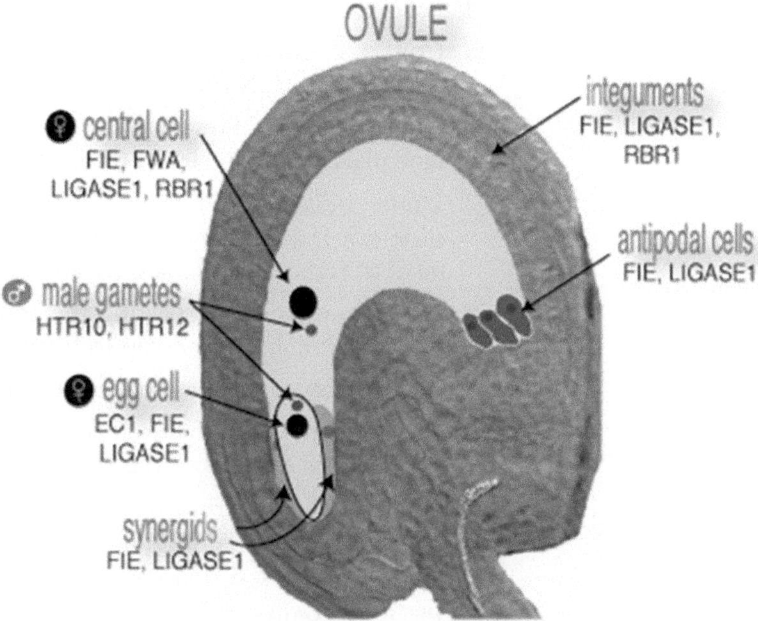


**Fig. 3** Combination of fluorescent markers to perform live imaging of double fertilization. Presented fluorescent markers are highly expressed and facilitate live imaging of events leading to double fertilization. HTR10 and EC1 markers are tagged with a red fluorescent protein (RFP) tag. All other markers are expressing a green fluorescent protein (GFP) tag. References corresponding to the mentioned genes and transgenic marker lines: *EC1* [22, 26]; *FWA* [27]; *FIE* [15, 34]; *HTR10* [15]; *HTR12* [16, 17]; *LIGASE1* [25, 28]; *RBR1* [15, 35]

the endosperm. A precise time course study of double fertilization events on fixed materials established that the first pollen tube discharge takes place within 5 hours after pollination (HAP) [32]. All the embryo sacs have reached that stage by 9 HAP [32]. Fusion of male and female nuclei (karyogamy) occurs 11–12 HAP. Within 1 h (13 HAP), the first division of endosperm is detectable. This description obtained on fixed tissues was recently confirmed for the first time using live imaging confocal microscopy using fluorescent markers for male and female gametes [12, 15]. Live imaging of double fertilization has been further studied using additional combinations of fluorescent male and female markers [7, 22, 25, 33]. Fluorescent marker lines recommended for live imaging of double fertilization are summarized in Fig. 3.

1. Emasculate the last closed flower containing the selected fluorescent marker expressed in cells of the embryo sac (*see* **Note 6**).
2. Remove any nearby open flowers.
3. 24 h after emasculation, hand-pollinate emasculated pistils with pollen from a freshly open flower expressing a compatible fluorescent marker (*see* **Notes 6** and 7).

4. Label each cross by tying a ribbon of color tape around the stem.
5. Collect a pollinated flower by the peduncle with a fine forceps No. 3 (DUMONT).
6. Affix the selected flower on the ribbon of double-sided adhesive tape placed on a square Petri dish.
7. Under a dissecting microscope fitted with a cold light source, slide down a bent hypodermic needle along each side of the pistil to remove the sepals, petals, and stamens.
8. Slit open the pistil by passing the bent hypodermic needle just under the replum from top to bottom (*see* **Note 4**).
9. Cut the top of the pistil where stigmatic papillae would form.
10. Incise gently one of the two valves of the pistil placing the hypodermic needle between the valve and the row of ovules.
11. Collect with a superfine forceps No. 5 (DUMONT) the septum with files of ovules holding firmly the remaining valve with a fine forceps No. 3 (DUMONT).
12. Place the septum onto a glass slide containing few drops of MS medium (1/4, pH 5.8) supplemented with 0.1 % gelrite (*see* **Note 5**).
13. Cover the sample with a coverslip.

### 3.5 Live Imaging of Female Gametophyte Development and Events Leading to Double Fertilization

Laser scanning confocal microscopy is recommended to image within thick materials like ovules. Dynamics of compatible fluorescence markers (e.g., RFP versus GFP) is observed in the resulting cross from pollen carrying a fluorescent sperm cell marker and ovules carrying a fluorescent cell-type-specific or embryo sac marker. To image events leading to double fertilization, the experimenter dissects the pollinated pistils at time points as detailed [15, 32]. Using laser scanning confocal microscopy (Zeiss LSM-510), fluorescence is usually acquired sequentially for GFP with selective settings for GFP detection (excitation, 488 nm; emission, band-pass 510–550 nm) and for RFP (excitation, 543 nm; emission, band-pass 560–615 nm). These settings generate autofluorescence that outline the contours of the mature embryo sac. Additional settings can be used to acquire autofluorescence of other ovule compartments. To highlight the central cell cytoplasm in the mature embryo sac and surrounding integument cells, autofluorescence is simultaneously collected with selective settings for GFP detection (excitation, 488 nm; emission, 510–550 nm) and non-specific settings for autofluorescence detection (excitation, 488 nm; emission, long-pass 560 nm). Autofluorescence generally becomes fainter as earlier stages of female gametophyte development are studied. As a substitute, the experimenter can in parallel collect images using Nomarski optics to visualize overall structure of ovules and embryo sacs.

## 4 Notes

1. Distribute under a laminar hood the sterilized medium into aliquots in sterile Eppendorf tubes. Store at room temperature.
2. To obtain healthy tricellular pollen grains, collect a freshly opened flower containing dehiscent anthers. At later stages of flower development pollen, grains become desiccated. To obtain earlier stages of male gametogenesis, collect younger flowers from the same inflorescence.
3. To obtain healthy mature female gametophyte, collect a freshly opened flower. Collect younger flowers on the same inflorescence to obtain earlier stages of female gametogenesis.
4. The flat part of the bent needle must be kept parallel to the surface of the petri dish to avoid damaging the ovules.
5. Use an aliquot of sterilized MS medium (*see* **Note 1**). Aliquots are usually rapidly contaminated upon use and should be discarded at the end of each experiment.
6. Depending on the skills of the experimenter, emasculation can be performed by eye or assisted with a magnifying device.
7. For details and tips on crosses, see ref. 36.

## Acknowledgments

The author would like to thank Fred Berger for his advice on imaging reproductive structures and Daniel Grimanelli for critical reading of the manuscript.

### References

1. Yadegari R, Drews GN (2004) Female gametophyte development. Plant Cell 16:S133–141
2. Borg M, Brownfield L, Twell D (2009) Male gametophyte development: a molecular perspective. J Exp Bot 60:1465–1478
3. Ma H, Sundaresan V (2010) Development of flowering plant gametophytes. Curr Top Dev Biol 91:379–412
4. Yang WC, Shi DQ, Chen YH (2010) Female gametophyte development in flowering plants. Annu Rev Plant Biol 61:89–108
5. Faure JE, Dumas C (2001) Fertilization in flowering plants. New approaches for an old story. Plant Physiol 125:102–104
6. Berger F (2011) Imaging fertilization in flowering plants, not so abominable after all. J Exp Bot 62:1651–1658
7. Sprunck S (2010) Let's get physical: gamete interaction in flowering plants. Biochem Soc Trans 38:635–640
8. Borges F, Gomes G, Gardner R, Moreno N, McCormick S, Feijo JA, Becker JD (2008) Comparative transcriptomics of Arabidopsis sperm cells. Plant Physiol 148:1168–1181
9. Wuest SE, Vijverberg K, Schmidt A, Weiss M, Gheyselinck J, Lohr M, Wellmer F, Rahnenfuhrer J, von Mering C, Grossniklaus U (2010) Arabidopsis female gametophyte gene expression map reveals similarities between plant and animal gametes. Curr Biol 20:506–512
10. Figueroa DM, Bass HW (2010) A historical and modern perspective on plant cytogenetics. Brief Funct Genomics 9:95–102

11. Ronceret A, Pawlowski WP (2010) Chromosome dynamics in meiotic prophase I in plants. Cytogenet Genome Res 129:173–183
12. Berger F, Hamamura Y, Ingouff M, Higashiyama T (2008) Double fertilization—caught in the act. Trends Plant Sci 13:437–443
13. Rotman N, Durbarry A, Wardle A, Yang WC, Chaboud A, Faure JE, Berger F, Twell D (2005) A novel class of MYB factors controls sperm-cell formation in plants. Curr Biol 15:244–248
14. Strompen G, Dettmer J, Stierhof YD, Schumacher K, Jurgens G, Mayer U (2005) Arabidopsis vacuolar H-ATPase subunit E isoform 1 is required for Golgi organization and vacuole function in embryogenesis. Plant J 41:125–132
15. Kim HJ, Oh SA, Brownfield L, Hong SH, Ryu H, Hwang I, Twell D, Nam HG (2008) Control of plant germline proliferation by SCF(FBL17) degradation of cell cycle inhibitors. Nature 455:1134–1137
16. Ingouff M, Hamamura Y, Gourgues M, Higashiyama T, Berger F (2007) Distinct dynamics of HISTONE3 variants between the two fertilization products in plants. Curr Biol 17:1032–1037
17. Fang Y, Spector DL (2005) Centromere positioning and dynamics in living Arabidopsis plants. Mol Biol Cell 16:5710–5718
18. Portereiko MF, Lloyd A, Steffen JG, Punwani JA, Otsuga D, Drews GN (2006) *AGL80* is required for central cell and endosperm development in Arabidopsis. Plant Cell 18:1862–1872
19. Jullien PE, Mosquna A, Ingouff M, Sakata T, Ohad N, Berger F (2008) Retinoblastoma and its binding partner *MSI1* control imprinting in Arabidopsis. PLoS Biol 6:e194
20. Johnson-Brousseau SA, McCormick S (2004) A compendium of methods useful for characterizing Arabidopsis pollen mutants and gametophytically-expressed genes. Plant J 39:761–775
21. Chen Z, Tan JL, Ingouff M, Sundaresan V, Berger F (2008) Chromatin assembly factor 1 regulates the cell cycle but not cell fate during male gametogenesis in *Arabidopsis thaliana*. Development 135:65–73
22. Aw SJ, Hamamura Y, Chen Z, Schnittger A, Berger F (2010) Sperm entry is sufficient to trigger division of the central cell but the paternal genome is required for endosperm development in Arabidopsis. Development 137:2683–2690
23. Wang D, Zhang C, Hearn DJ, Kang IH, Punwani JA, Skaggs MI, Drews GN, Schumaker KS, Yadegari R (2010) Identification of transcription-factor genes expressed in the Arabidopsis female gametophyte. BMC Plant Biol 10:e110
24. Steffen JG, Kang IH, Macfarlane J, Drews GN (2007) Identification of genes expressed in the Arabidopsis female gametophyte. Plant J 51:281–292
25. Ingouff M, Sakata T, Li J, Sprunck S, Dresselhaus T, Berger F (2009) The two male gametes share equal ability to fertilize the egg cell in *Arabidopsis thaliana*. Curr Biol 19:R19–20
26. Wang D, Tyson MD, Jackson SS, Yadegari R (2006) Partially redundant functions of two SET-domain polycomb-group proteins in controlling initiation of seed development in Arabidopsis. Proc Natl Acad Sci U S A 103:13244–13249
27. Kinoshita T, Miura A, Choi Y, Kinoshita Y, Cao X, Jacobsen SE, Fischer RL, Kakutani T (2004) One-way control of FWA imprinting in Arabidopsis endosperm by DNA methylation. Science 303:521–523
28. Andreuzza S, Li J, Guitton AE, Faure JE, Casanova S, Park JS, Choi Y, Chen Z, Berger F (2010) DNA LIGASE I exerts a maternal effect on seed development in *Arabidopsis thaliana*. Development 137:73–81
29. Sandaklie-Nikolova L, Palanivelu R, King EJ, Copenhaver GP, Drews GN (2007) Synergid cell death in Arabidopsis is triggered following direct interaction with the pollen tube. Plant Physiol 144:1753–1762
30. Bemer M, Heijmans K, Airoldi C, Davies B, Angenent GC (2010) An atlas of type I MADS box gene expression during female gametophyte and seed development in Arabidopsis. Plant Physiol 154:287–300
31. Christensen CA, King EJ, Jordan JR, Drews GN (1997) Megagametogenesis in Arabidopsis wild type and the *Gf* mutant. Sex Plant Reprod 10:49–64
32. Faure JE, Rotman N, Fortune P, Dumas C (2002) Fertilization in *Arabidopsis thaliana* wild type: developmental stages and time course. Plant J 30:481–488
33. Maruyama D, Endo T, Nishikawa S (2010) BiP-mediated polar nuclei fusion is essential for the regulation of endosperm nuclei proliferation in *Arabidopsis thaliana*. Proc Natl Acad Sci U S A 107:1684–1689
34. Yadegari R, Kinoshita T, Lotan O, Cohen G, Katz A, Choi Y, Nakashima K, Harada JJ, Goldberg RB, Fischer RL, Ohad N (2000) Mutations in the FIE and MEA genes that encode interacting polycomb proteins cause parent-of-origin effects on seed development by distinct mechanisms. Plant Cell 12:2367–2382
35. Ingouff M, Jullien PE, Berger F (2006) The female gametophyte and the endosperm control cell proliferation and differentiation of the seed coat in Arabidopsis. Plant Cell 18:3491–3501
36. Weigel D, Glazebrook J (2002) Arabidopsis: a laboratory manual. Cold Spring Harbor Laboratory Press, New York

# Chapter 9

# Genome-Wide Analysis of DNA Methylation in *Arabidopsis* Using MeDIP-Chip

**Sandra Cortijo, René Wardenaar, Maria Colomé-Tatché, Frank Johannes, and Vincent Colot**

## Abstract

DNA methylation is an epigenetic mark that is essential for preserving genome integrity and normal development in plants and mammals. Although this modification may serve a variety of purposes, it is best known for its role in stable transcriptional silencing of transposable elements and epigenetic regulation of some genes. In addition, it is increasingly recognized that alterations in DNA methylation patterns can sometimes be inherited across multiple generations and thus are a source of heritable phenotypic variation that is independent of any DNA sequence changes. With the advent of genomics, it is now possible to analyze DNA methylation genome-wide with high precision, which is a prerequisite for understanding fully the various functions and phenotypic impact of this modification. Indeed, several so-called epigenomic mapping methods have been developed for the analysis of DNA methylation. Among these, immunoprecipitation of methylated DNA followed by hybridization to genome tiling arrays (MeDIP-chip) arguably offers a reasonable compromise between cost, ease of implementation, and sensitivity to date. Here we describe the application of this method, from DNA extraction to data analysis, to the study of DNA methylation genome-wide in *Arabidopsis*.

**Key words** DNA methylation, 5-Methylcytosine (5mC), MeDIP, Tiling array, Epigenetic variation

## 1 Introduction

In eukaryotes, DNA methylation almost exclusively affects cytosines (5-methylcytosines). Once established, this modification can be maintained over numerous cell divisions and even across generations in some instances. However, it remains unclear to what extent differences in DNA methylation can be stably inherited and this question is the subject of intense studies. This is especially true in *Arabidopsis*, where epigenetic recombinant inbred lines (epiRILs) have been derived from parents with few differences in DNA sequence but contrasted DNA methylation profiles [1, 2]. One such population of epiRILs has been epigenotyped [3] in order to assess the stability of parental DNA methylation differences and

Charles Spillane and Peter C. McKeown (eds.), *Plant Epigenetics and Epigenomics: Methods and Protocols*, Methods in Molecular Biology, vol. 1112, DOI 10.1007/978-1-62703-773-0_9, © Springer Science+Business Media New York 2014

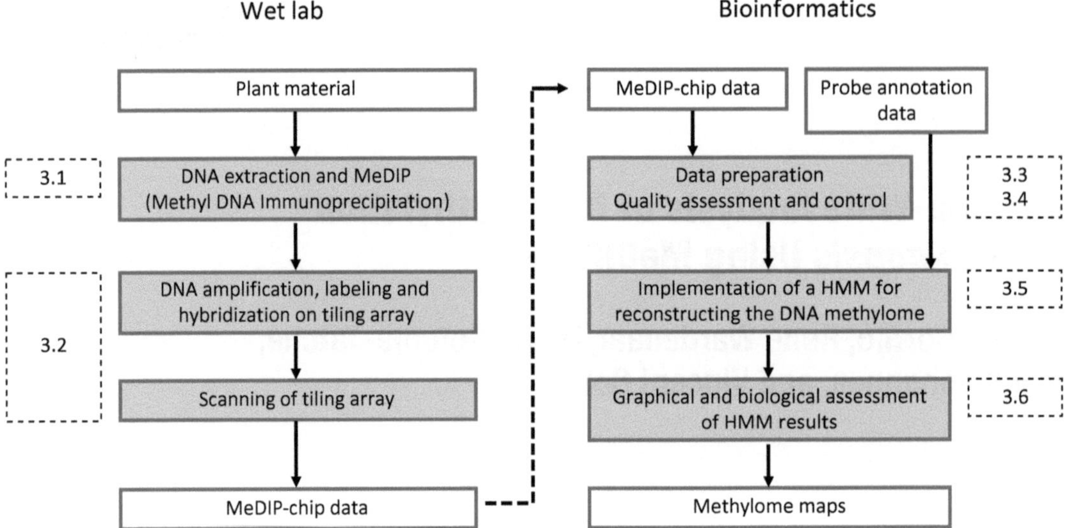


**Fig. 1** Flowchart for the reconstruction of methylome maps (Subheading 3)

their impact on several complex traits. Here, we describe the methyl DNA immunoprecipitation (MeDIP)-chip protocol used to reconstruct the DNA methylome maps, starting from the extraction of DNA to analysis of the hybridization data using Hidden Markov Models (HMM; *see* flow chart in Fig. 1). Subheading 2 lists the materials needed for the "wet" part as well as the software and data used for analysis. Subheading 3 describes step by step the MeDIP-chip experiment (Subheadings 3.1 and 3.2) and the analysis of hybridization data, starting from data preparation (Subheading 3.3), then quality assessment and control (Subheading 3.4), implementation of a HMM for reconstructing DNA methylome maps (Subheading 3.5) and graphical and biological assessment of HMM results (Subheading 3.6).

## 2 Materials

### *2.1 DNA Extraction and Methyl DNA Immunoprecipitation*

1. DNA extraction with DNeasy plant Maxi kit (Qiagen, Catalogue N° 68163).
2. 1.5 mL Siliconized tubes: Clear-view™ Snap-Cap microtubes, siliconized (Sigma, Catalogue N° T4816-250EA).
3. Sonicator: Bioruptor (Diagenode, Catalogue N° UCD-200).
4. Buffer 1: 13.3 mM Tris–HCl pH 7.5, 667 mM NaCl, 1.3 mM EDTA.
5. Monoclonal antibody against 5mC (Diagenode, Catalogue N° MAb-006-500).
6. Rotating wheel.

7. Magnetic beads: M280 Dynabeads (Invitrogen, Catalogue # 112-01D).
8. Buffer 2: 10 mM Tris–HCl pH 7.5, 500 mM NaCl, 1 mM EDTA.
9. Buffer 3: 30 mM Tris–HCl pH 8.0.
10. Proteinase K: 20 μg/μL (NEB, Catalogue N° P8102S).
11. Phenol/chloroform/IAA (25:24:1, pH 8.0) and Chloroform/IAA: (24:1).
12. Glycogen azure: 20 μg/μL, resuspended in water (Sigma, Catalogue N° G5510-1G).
13. NaOAc: 3 M, pH 5.2.
14. MinElute Reaction Cleanup Kit (Qiagen, Catalogue N° 28204).
15. PicoGreen: Quant-it PicoGreen dsDNA reagent (Invitrogen, Catalogue # P7581) diluted to 0.5 % in TE, pH 8.

### 2.2 DNA Amplification, Labeling, and Hybridization on Tiling Array

1. WGA2 kit (Sigma, Catalogue # WGA2-50RXN).
2. QIAquick PCR Purification Kit (Qiagen, Catalogue N° 28104).
3. Dual Color DNA labeling kit (NimbleGen, Catalogue N° 06370250001).
4. Hybridization and wash buffer kits (NimbleGen, Catalogue N° 05583683001 and 05584507001).
5. Scanner: High-Resolution (2 μm) Microarray Scanner (Agilent, Catalogue N° G2565CA).
6. NimbleScan software (NimbleGen).

### 2.3 Software Requirements

This protocol requires R for the analysis of the MeDIP-chip data. R is a command line-based software environment for statistical computing and graphics. It can be freely downloaded at http://www.r-project.org and installed on all three main operating systems (Windows, Unix/Linux, and Mac). Instructions about installation and tutorials can be obtained from the same website. R is extensively used among biostaticians due to the availability of statistical packages for the analysis of a broad spectrum of biological data. In addition to R, we also recommend downloading a text editor with syntax highlighting (e.g., Notepad++). Programming mistakes are more easily detected when using a text editor. All the code lines and functions are highlighted throughout the chapter in courier font. The HMM is implemented in C++. An electronic version of the R code presented in this chapter and the HMM software are freely available at the following URL: http://www.johanneslab.org/publications. This chapter does not show the code for generating the figures. This code can, however, be downloaded from the same URL.

**Table 1**
**Format methylation data**

| PROBE_ID | REP1_INPUT_RED | REP2_INPUT_RED | REP3_INPUT_RED |
|---|---|---|---|
| CHR01FS000000061 | 778.53 | 2534.67 | 1033.31 |
| CHR01FS000000212 | 2366.51 | 2756.02 | 1333.69 |
| CHR01FS000000382 | 4028.27 | 7776.75 | 3201.88 |
| CHR01FS000000507 | 13685.61 | 15014.29 | 8556.37 |
| CHR01FS000000707 | 1565.45 | 2626.51 | 1187.04 |

### 2.4 Dataset

The protocol was implemented for the efficient and cost-effective genome-wide study of DNA methylation of a large number of *Arabidopsis* lines. The dataset used to illustrate this protocol can be downloaded from the above URL and consists of six files that contain the measured signal intensities (IP and INPUT) for one wild type line (Columbia accession, Col-0), probe annotation, conservation scores for probes, and an example of an array with a hybridization artefact.

#### 2.4.1 Methylation Data

The methylation data should be tab-delimited and have the format shown in Table 1. The first column of the file should contain the probe identifier and the remaining column (or columns when replicates are available) should contain the measured probe intensities. The IP and INPUT files should have the same tab-delimited format.

#### 2.4.2 Hybridization Artefact Data

For illustrative purposes, we also show an example of a hybridization artefact (Fig. 2a). This file should also be tab-delimited and have the format shown in Table 2. The first column should again contain the probe identifier, the second and third column should contain the location of the probe on the array ($x$ and $y$ position on the array) and the fourth column (PM) should contain the measured probe intensity (IP or INPUT signal).

#### 2.4.3 Conservation Score Data

The conservation score of a probe indicates the uniqueness of the probe sequence (not all probe sequences are unique). This score was obtained by performing a BLAST search. Scores are percentage of identity with the second best hit (the first hit is the location in the genome for which the probe was designed). Probes can be visualized at http://epigara.biologie.ens.fr/index.html. The conservation score data should have the tab-delimited format shown in Table 3.

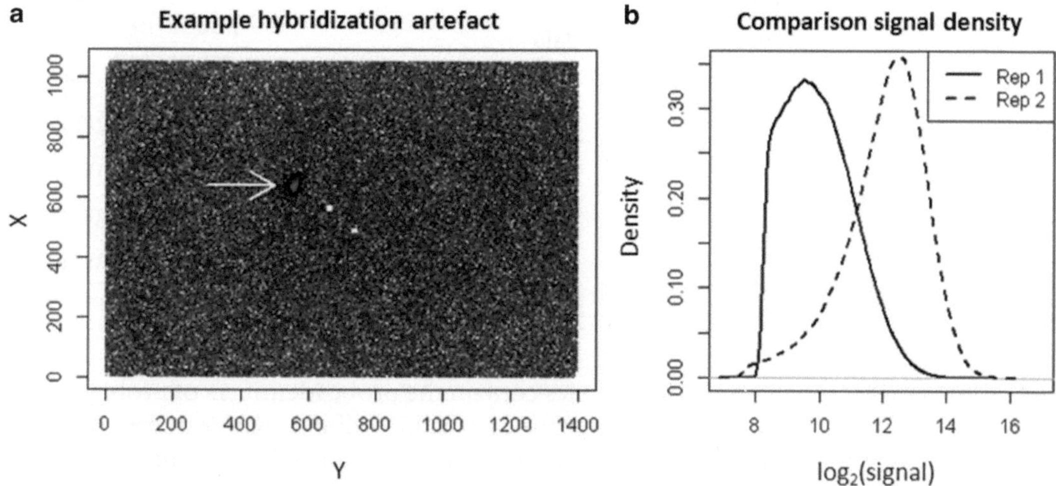


**Fig. 2** Quality of the overall hybridization experiment. (**a**) The *arrow* points to an unwanted spatial artefact on the tiling array. One could consider excluding the relevant probes or discarding the tiling array entirely. (**b**) Shown is the signal density distribution of the Cy3 INPUT channel for two replicates. One of the individuals (*solid line*) shows a steep increase in the lower signal range suggesting that an insufficient amount of DNA was hybridized to the tiling array. The signal distribution of the other replicate (*dashed line*) is normal. The bulk of the data is located in the center of the detection range indicating that the right amount of DNA was hybridized to the tiling array

**Table 2**
**Format hybridization artefact data**

| PROBE_ID | X | Y | PM |
|---|---|---|---|
| CHR01FS000000061 | 327 | 1335 | 3219.96 |
| CHR01FS000000212 | 191 | 1257 | 4840.31 |
| CHR01FS000000382 | 826 | 34 | 8668.02 |
| CHR01FS000000507 | 731 | 529 | 19781.76 |
| CHR01FS000000707 | 624 | 562 | 1195.29 |

**Table 3**
**Format conservation score data**

| PROBE_ID | Score |
|---|---|
| CHR01FS000000061 | 73 |
| CHR01FS000000212 | 56 |
| CHR01FS000000382 | 64 |
| CHR01FS000000507 | 62 |
| CHR01FS000000707 | 74 |

**Table 4**
**Format annotation data**

| PROBE_ID |
|---|
| CHR01FS000004351 |
| CHR01FS000005311 |
| CHR01FS000007129 |
| CHR01FS000007479 |
| CHR01FS000007814 |

#### 2.4.4 Annotation Data

The annotation files contain the probe identifiers of probes that are located within introns of protein-coding genes or transposons. The annotation data should only contain one column with probe identifiers as shown in Table 4.

## 3 Methods

### 3.1 DNA Extraction and MeDIP

1. Extract DNA from plant material (1–2 g fresh weight, we use aerial parts of 3-week-old plants grown under long day conditions) with Qiagen DNeasy plant Maxi kit. 1.8 μg of DNA is needed for this protocol (includes sonication test and INPUT and IP fractions).
2. Quantify DNA and place 1.8 μg in a final volume of 180 μL (complete with water if necessary) in 1.5 mL siliconized Eppendorf tubes. Set aside 2 μL (corresponding to 20 ng of DNA) for sonication control. Sonicate the remaining 178 μL using seven cycles of 30 s ON/30 s OFF. Note that all six positions within the sonicator need to be filled, with an equal volume of water (178 μL) put in each tube. Place all six tubes in an ice bucket and add ice to the sonicator bath to cool it off. Repeat sonication once (14 cycles in total). Keep 13 μL to test sonication (sonicated fraction).
3. Run non-sonicated (2 μL) and sonicated (13 μL) samples side by side in 1.5 % 1× TAE gel. A smear should be visible between 100 and 600 bp (with maximum intensity around 300 bp) after sonication.
4. Keep 15 μL to serve as INPUT (150 ng). Use the remaining 150 μL of sonicated DNA (1.5 μg) for IP.
5. Add 450 μL of buffer 1 to IP fraction (total volume of 600 μL). Incubate 10 min at 95 °C to denature DNA (this is critical as the antibody only recognizes 5mC on single-stranded DNA!) and let sit on ice for 2 min. Add 5 μL of 1 μg/μL anti-5mC antibody to denatured IP fraction. Close

tubes, wrap with parafilm (siliconized tubes tend to leak) and incubate overnight at 4 °C with gentle agitation (we use a rotating wheel, with a 45° inclination, 8 rpm).

6. Use 40 μL of magnetic beads per MeDIP. Prepare a tube with the total amount of beads required for the number of MeDIP performed. Wash the beads three times with 1 mL of buffer 2 (*see* **Note 1**) and resuspend one last time with buffer 2 in the starting volume. Put 40 μL of washed beads, making sure that they are well resuspended by pipetting up and down the slurry several times, into each MeDIP tube. Put on rotating wheel for 4 h at 4 °C with gentle agitation (45° inclination, 8 rpm).
7. Put IP samples on the Dynabeads rack (magnetic rack). Collect supernatant in a new 2 mL Eppendorf tube (supernatant fraction). Add 300 μL of buffer 2 to IP tube. Agitate briefly by hand, and place for 10 min at room temperature on the rotating wheel with gentle agitation (45° inclination, 8 rpm). Put back on the Dynabeads rack and add first wash to supernatant fraction. Perform three more washes, each time with 600 μL of buffer 2. Discard washes.
8. Add 300 μL of buffer 3 to the IP pellet after last wash and transfer IP and supernatant fractions to 1.5 mL and 2 mL tubes, respectively (*see* **Note 2**). Add 7 μL of Proteinase K to elute. Incubate 1 h at 42 °C, with occasional shaking.
9. Add one volume of phenol/chloroform/IAA to the IP and supernatant fractions (300 and 900 μL, respectively). Vortex and centrifuge 5 min at 14,000 × $g$ at room temperature. Place aqueous phase (top phase) in a new tube. Add one volume of chloroform/IAA to aqueous phase. Vortex and centrifuge 5 min at 14,000 × $g$ at room temperature. Place aqueous phase in a new tube.
10. To precipitate DNA, add 1 μL of glycogen azure, 1:10 volume of NaOAc, and one volume of isopropanol to the IP and supernatant fractions (30 and 90 μL for NaOAc and 300 and 900 μL for isopropanol in IP and supernatant fraction, respectively). Vortex between addition of each component. Keep at −20°C for at least 1 h or overnight. Centrifuge 30 min at room temperature at max speed (>13,000 × $g$). Discard supernatant and add 500 μL of ethanol 70 %. Mix and centrifuge for 20 min at room temperature at max speed (>13,000 × $g$). Discard the supernatant and dry DNA pellets by leaving the tubes open on the bench for ~30 min. Resuspend all DNA pellets in 40 μL of TE, pH 8.0, and add 25 μL to INPUT fraction.
11. Perform quantitative PCR on the three fractions (IP, supernatant, and INPUT) with known positive and negative controls before proceeding with purification, labeling, and hybridization

to tiling array. Note that for wild type *Arabidopsis* (Columbia accession), approximately 10–20 % of the genome should be immunoprecipitated with the anti-5mC antibody for DNA extracted from aerial or root parts.

12. DNA should be cleaned one last time using the MinElute kit (*see* **Note 3**). Expect 30 % loss of DNA.
13. DNA concentration is checked with Nanodrop 3300. Add 2 μL of diluted PicoGreen at 0.5 % to 2 μL of DNA and quantify this mix using function "dsDNA PicoGreen® dye" in "Nucleic Acid Quantitation" (*see* **Note 4**).

### 3.2 DNA Amplification, Labeling, and Hybridization on Tiling Array

1. Use 10 ng of IP and 50 ng of INPUT fractions for amplification with the WGA2 kit. Start from the "Library preparation" step of the protocol, as there is no need for the DNA fragmentation step.
2. Purification of the amplification products is carried out using QIAquick PCR Purification Kit. Quantify and run in a 1.5 % agarose TAE 1× gel. This should produce a smear corresponding to the sonication smear (between 100 and 600 bp). Final yield fluctuates between 3 and 6 μg.
3. DNA labeling is carried out using the Dual Color DNA labeling kit, using 1 μg of amplified IP and INPUT DNA. Resuspend labeled DNA in 20 μL of water and quantify it, together with Cy3 and Cy5 using the "microarray function" of the Nanodrop 2000. One should expect 10–20 μg of DNA after labeling and 200–400 pmol of incorporated dye. Repeat labeling if DNA yield or incorporation levels are less than 5 μg or 100 pmol, respectively (*see* **Note 5**).
4. Differential hybridization is carried out using a NimbleGen 3x720K tiling array design (three identical chambers, design available on request) and following the manufacturer's instructions. Use 4 μg of each of the two labeled DNA samples (IP and INPUT) per chamber. Hybridization is in dye-swap (IP in red and INPUT in green for the first chamber and vice versa for the second chamber).
5. After washing, the NimbleGen 3x720K tiling array is scanned using a High-Resolution (2 μm) Microarray Scanner (Agilent). It is preferable to scan each chamber independently.
6. Grid alignment and pair files extraction are made using the NimbleScan software and following the manufacturer's instructions.

### 3.3 Data Preparation

Following the "wet lab" part, one is confronted with a substantial amount of data ready to be analyzed. Before we show how this can be achieved, we detail several data preparation steps. The following commands are used to import the data in the R workspace. The command `setwd()` sets the working directory, such that there is no need to define the complete pathname of your files. The command `head()` shows the first lines of the file.

```
> setwd("D:\\reconstruction_methylome_maps")
>  input_wt  <-  read.table(file="input_wild_type.
txt",
+ header=TRUE,sep="\t")
> ip_wt    <- read.table(file="ip_wild_type.txt",
+ header=TRUE,sep="\t")
>
> head(input_wt)
```

| | PROBE_ID | REP1_INPUT_RED | REP2_INPUT_RED | REP3_INPUT_RED |
|---|---|---|---|---|
| 1 | CHR01FS000000061 | 778.53 | 2534.67 | 1033.31 |
| 2 | CHR01FS000000212 | 2366.51 | 2756.02 | 1333.69 |
| 3 | CHR01FS000000382 | 4028.27 | 7776.75 | 3201.88 |
| 4 | CHR01FS000000507 | 13685.61 | 15014.29 | 8556.37 |
| 5 | CHR01FS000000707 | 1565.45 | 2626.51 | 1187.04 |
| 6 | CHR01FS000000827 | 5939.94 | 7285.02 | 3212.73 |

| | REP1_INPUT_GREEN | REP2_INPUT_GREEN | REP3_INPUT_GREEN |
|---|---|---|---|
| 1 | 408.61 | 2038.57 | 818.98 |
| 2 | 712.76 | 2019.65 | 649.84 |
| 3 | 1350.67 | 5406.18 | 2090.43 |
| 4 | 2980.53 | 9570.41 | 5614.20 |
| 5 | 611.33 | 2405.53 | 460.63 |
| 6 | 1162.24 | 4555.31 | 2311.96 |

The IP and INPUT data have the same format; hence, there is no need to show the first lines of both files. We convert the data to a logarithmic scale using the following commands:

```
> log2_ip_wt                <- log2(ip_wt[,2:7])
> log2_ip_wt                <-  data.frame(ip_wt[,1],
                            log2_ip_wt)
> names(log2_ip_wt)[1]      <- "PROBE_ID"
>
> log2_input_wt             <- log2(input_wt[,2:7])
> log2_input_wt             <- data.frame(input_wt[,1],
                            log2_input_wt)
> names(log2_input_wt)[1]   <- "PROBE_ID"
```

After log transformation, the datasets will have the same format only the signal intensities will be log transformed. In order to determine enrichment for DNA methylation, one has to calculate the intensity ratio of the IP and INPUT signal ($log_2$ ratios). The following commands are used to calculate the intensity ratios. The dye-swapped replicates have been treated separately in this case (i.e., $IP_{green}$ and $INPUT_{red}$ and vice versa). The IP and INPUT signals have also been averaged.

```
> wt_ip_green          <- (log2_ip_wt[,5]+log2_ip_wt
                          [,6]+log2_ip_wt[,7])/3
> wt_input_red         <-   (log2_input_wt[,2]+log2_
                          input_wt[,3]+
+ log2_input_wt[,4])/3
> wt_green_red         <- wt_ip_green-wt_input_red
> wt_green_red         <-                     data.
                          frame(log2_ip_wt[,1],wt_
                          green_red)
> names(wt_green_red) <- c("PROBE_ID","GREEN_RED")
>
> wt_ip_red            <- (log2_ip_wt[,2]+log2_ip_wt[,3]
                          +log2_ip_wt[,4])/3
> wt_input_green       <-          (log2_input_wt[,5]
                          +log2_input_wt[,6]+
+ log2_input_wt[,7])/3
> wt_red_green         <- wt_ip_red-wt_input_green
> wt_red_green         <- data.frame(log2_ip_wt[,1],
                          wt_red_green)
> names(wt_red_green) <- c("PROBE_ID","RED_GREEN")
```

After the calculation of the intensity ratios, the dye-swap signals can be calculated using the following code:

```
> wt_dye_swap          <-   (wt_green_red[,2]+wt_red_
                          green[,2])/2
> wt_dye_swap          <- data.frame(wt_green_red[,1],
                          wt_dye_swap)
> names(wt_dye_swap)   <- c("PROBE_ID","DYE_SWAP")
```

The dye-swap should account for possible dye bias in experiments. The data is now ready for subsequent analysis steps.

### 3.4 Quality Assessment and Control

Prior to array data analysis, we conduct detailed quality checks of each tiling array experiment. This quality assessment is necessary to ensure biologically meaningful results later on. If the data contains systematic hybridization artefacts or technical variation beyond a certain acceptable level, it is advisable to remove or to repeat the bad sample. We distinguish between two levels of quality assessment. The first level relates to the quality of the overall hybridization experiment and the second level to the quality of the individual probes.

#### 3.4.1 Quality of the Overall Hybridization Experiment

We start by evaluating the distribution (or spreading) of the DNA fragments over the tiling array. This can be achieved by visual inspection of the array image within each separate channel (Fig. 2a).

By design, the signals should be randomly distributed and show no systematic spatial patterns. Artefacts such as scratches and bright spots can be easily detected in this way. The following commands are used to import the data in the R workspace:

```
>hybr_artefact<-read.table(file="hybridization_
artefact.txt",
+ header=TRUE,sep="\t")
> head(hybr_artefact)
```

| | PROBE_ID | X | Y | PM |
|---|---|---|---|---|
| 1 | CHR01FS000000061 | 327 | 1335 | 3219.96 |
| 2 | CHR01FS000000212 | 191 | 1257 | 4840.31 |
| 3 | CHR01FS000000382 | 826 | 34 | 8668.02 |
| 4 | CHR01FS000000507 | 731 | 529 | 19781.76 |
| 5 | CHR01FS000000707 | 624 | 562 | 1195.29 |
| 6 | CHR01FS000000827 | 927 | 485 | 7460.27 |

Plotting the reconstructed array image involves log transformation of the measured signals (PM) and rescaling of the log transformed signal between 0 and 1 in order to convert the signal into RGB colors. The code for plotting the array image (Fig. 2a) can be found at the above URL (*see* end of Subheading 2.3).

We also evaluate whether a sufficient amount of DNA was hybridized to the array. This can be done by plotting the density of the signal of each separate channel (Fig. 2b). The detection range of the signal has a lower and upper bound. In the case of insufficient DNA, there will be a rapid increase of probe signals in the lower detection range. Conversely, in the case of too much DNA the signal distribution will become saturated in the upper detection range. Both scenarios can seriously compromise the sensitivity of the technology to capture biologically meaningful variation. To illustrate this, we plot the density of the input signal of two different arrays (Fig. 2b) using the plotting code that is provided as a text file (*see* end of Subheading 2.3).

#### 3.4.2 Quality of Individual Probes

The second quality assessment level is the quality of the probes. NimbleGen arrays are designed to minimize cross-hybridization as much as possible. However, given the large number of probes and near full genome coverage, it is difficult to exclude possible cross-hybridization events. Such events occur when nontarget sequences hybridize with probes on the array, leading to exaggerated signal intensities. It may therefore be desirable to identify probes, a priori, that have multiple similar or exact matches in the genome. We assess this by calculating the so-called conservation score. This score is obtained by performing a BLAST search. Scores are percentage of

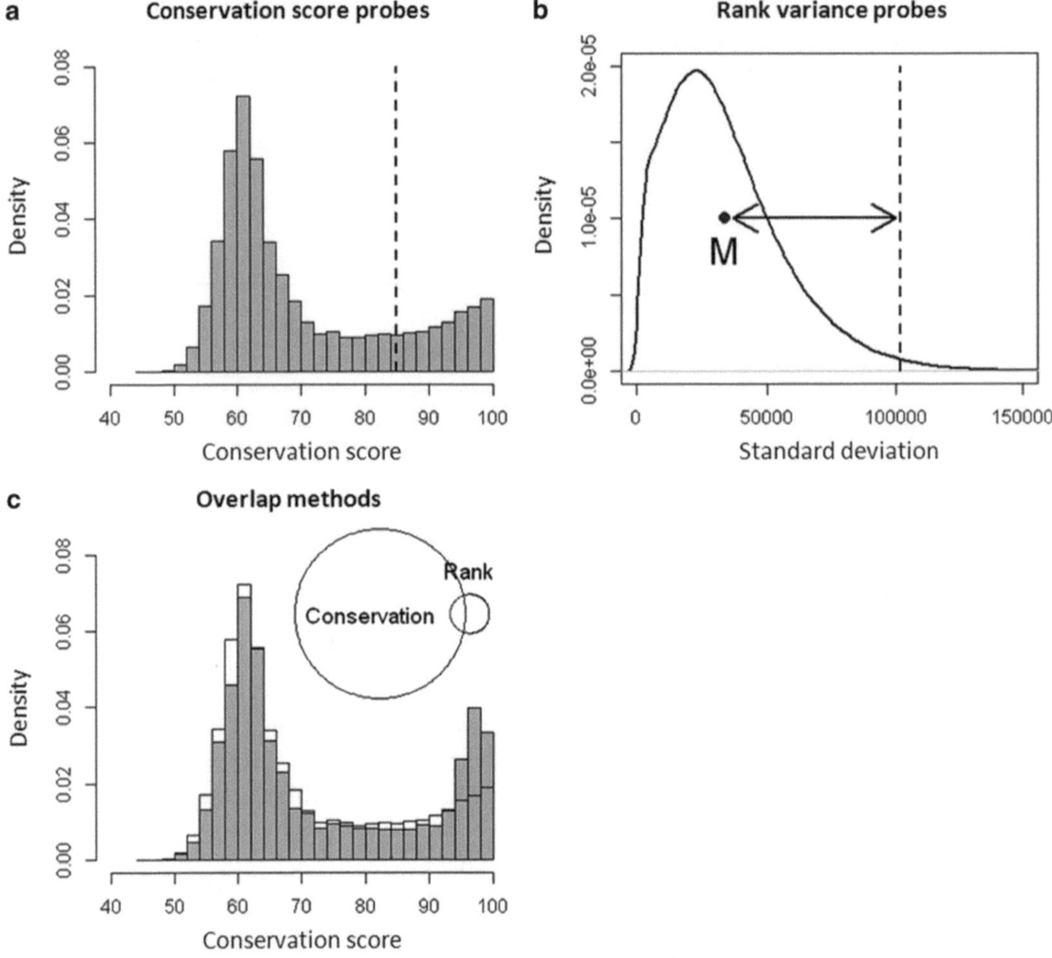


**Fig. 3** Quality of individual probes. (**a**) Density histogram of the conservation score of the probes. Probes with a conservation score higher than 85 have a high probability to cross-hybridize and are flagged (probes on the *right* of the *dashed line*). (**b**) The rank variance distribution of the probes. The rank variance is expressed as a standard deviation. Probes with an abnormal high rank variance are flagged (probes on the *right* of the *dashed line*). (**c**) Density histogram of the conservation score of the rank variance probes that were flagged (*gray*) on *top* of the conservation score of all probes (*transparent*). This picture indicates that the probes with a high rank variance also tend to have a high conservation score. The Venn diagram shows however that there is a poor overlap between probes that are flagged with the two methods

identity with the second best hit (the best hit is the location on the genome for which the probe was designed). We decided to flag probes that have a conservation score higher than 85 (Fig. 3a). For simplicity, we here provide a complete dataset with conservation scores already assigned to each probe.

This data can be inputted as follows:

```
> cons_score_probes  <-  read.table(file="conservation_
  score.txt",
+ header=TRUE,sep="\t")
> head(cons_score_probes)
```

| | PROBE_ID | SCORE |
|---|---|---|
| 1 | CHR01FS000000061 | 73 |
| 2 | CHR01FS000000212 | 56 |
| 3 | CHR01FS000000382 | 64 |
| 4 | CHR01FS000000507 | 62 |
| 5 | CHR01FS000000707 | 74 |
| 6 | CHR01FS000000827 | 66 |

One can use the plotting code which is provided as a text file to plot the density histogram of the conservation scores of the probes as shown in Fig. 3a.

In addition to the above a priori screening of potential cross-hybridizing probes, we utilize another quality criterion, which involves assessing the consistency of probe signals for the INPUT across biological or technical replicates (provided they are available). To do this, we identify a probe's signal rank in the overall array signal distribution of one replicate array and compare it to its rank in the distribution of the other arrays. Inconsistent probe signals will show large variation in ranks and should be treated with caution. If we consider the three dye-swapped biological replicates ($3 \times 2$ arrays) of the Col-0 accessions, there are six rank values for each probe, and we can calculate its rank variance. Doing this for each probe on the array yields a rank variance distribution, which can be used to spot outlying probes (Fig. 3b). For example, we may want to consider excluding or flagging probes with a rank variance of more than 3 standard deviations from the mean (Fig. 3b).

We use the following code to determine the rank and the rank variance of the probes as well as the three standard deviation cutoff:

```
> probe_rank        <-     apply(log2_input_wt[,2:7],
                  MARGIN=2,rank)
> determine_rank_var <- function(x){
+    mean_val       <- mean(x)
+    mean_dif       <- abs(x-mean_val)
+    extreme        <- which(mean_dif == max(mean_dif))
+    sd_ext         <- sd(x[-extreme])
+    return(sd_ext)
+ }
```

```
> rank_var          <-    apply(probe_rank,MARGIN=1,
                    determine_rank_var)
> rank_var          <-  data.frame(log2_input_wt[,1],
                    rank_var)
> names(rank_var)   <- c("PROBE_ID","SD")
> mean_var          <- mean(rank_var[,2])
> sd_var            <- sd(rank_var[,2])
> sd_cutoff         <- mean_var+(3*sd_var)
```

The plot of the rank variances (Fig. 3b) can be generated using the plotting code that is provided as a text file (*see* end of Subheading 2.3).

We find that the use of conservation scores and probe rank variance provides a fairly comprehensive assessment of probe quality. That these two criteria are not redundant is reflected in the limited overlap of identified low quality probes (Fig. 3c). We determine this overlap using the following code:

```
> lowq_pr_rank  <- rank_var[which(rank_var[,2]  > sd_
                cutoff),1]
> lowq_pr_cons  <- cons_score_probes[which(cons_score_
                probes[,2] > 85),1]
> lowq_probes   <- union(lowq_pr_rank,lowq_pr_cons)
> num_rank      <-  length(setdiff(lowq_pr_rank,lowq_
                pr_cons)) #Only rank
> num_cons      <-  length(setdiff(lowq_pr_cons,lowq_
                pr_rank)) #Only cons
> num_overlap   <- length(intersect(lowq_pr_rank,lowq_
                pr_cons))
> lowq_rows     <- which(wt_green_red[,1] %in% lowq_
                probes)
```

Finally, we plot the overlap between the two methods (Fig. 3c) using the plotting code.

### 3.4.3 The Effect of Removing Low Quality Probes

The removal of low quality probes has a visible impact on the overall signal distribution. To see this, we plot the relative (or ratio) signal of the IP and the INPUT channel in Fig. 4 on a $\log_2$ scale (*see* file with plotting code). High signals are typically an indication of increased IP hybridization events relative to the total (INPUT) DNA, thus indexing methylated DNA sequences. We find that most low quality scores fall in the upper signal range, suggesting that true binding events are partially confounded with cross-hybridization events. This is consistent with the observation, in *Arabidopsis*, that DNA methylation primarily occurs in CG-rich repeat elements [3, 4], which have a high cross-hybridization potential. For all subsequent analysis, we decided to keep (but flag) low quality probes in the dataset. However, one may also choose to exclude them at this stage.

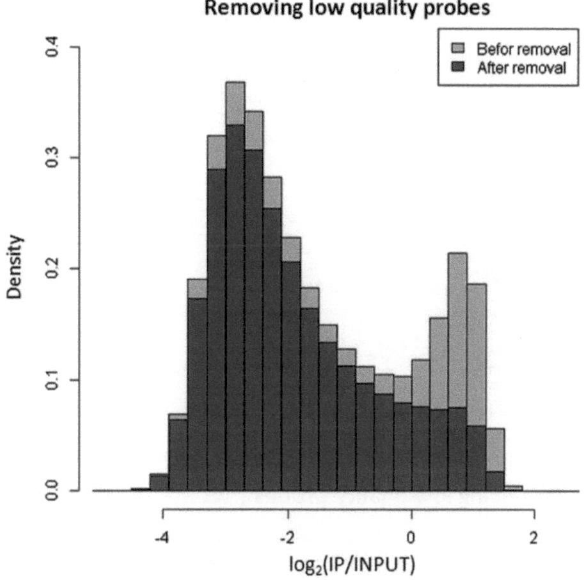


**Fig. 4** Effect of removing low quality probe signals from the overall $\log_2$(IP/INPUT) signal distribution. Most low quality signals fall in the upper range of the distribution, suggesting that true binding events are partially confounded with cross-hybridization events

### 3.5 Implementation of a Hidden Markov Model for Reconstructing the DNA Methylome

The above-mentioned log2 transformed IP/INPUT signal ratio is the typical starting point for data analysis. If data from several replicates is available, as in our case, the probe signals can simply be averaged across replicates. We view this distribution (*see* Fig. 4) as a mixture of three partially overlapping components [6]. The right component corresponds to enriched probes (i.e., indexing methylated sequences), the left component to non-enriched probes (i.e., indexing unmethylated sequences), and the middle component to intermediately enriched probes (i.e., indexing intermediately methylated sequences). To illustrate that this mixture view is consistent with the underlying biology, we highlight the probe signals corresponding to annotated transposable elements, which are usually methylated in *Arabidopsis* (Fig. 5a, solid line; [4, 5]). Similarly, as an example of usually unmethylated sequences, we highlight the signal of annotated introns of protein-coding genes (Fig. 5a, dashed line; [4, 5]).

The following commands are used to import the probe annotation data. These files simply contain the probe identifiers of probes that match with introns or transposons.

```
> p_id_intron <- read.table(file="intron_probes.txt",
+ header=TRUE,sep="\t")
> p_id_transp <- read.table(file="transposon_probes.txt",
+ header=TRUE,sep="\t")
    > head(p_id_intron)
    PROBE_ID
1 CHR01FS000004351
2 CHR01FS000005311
```

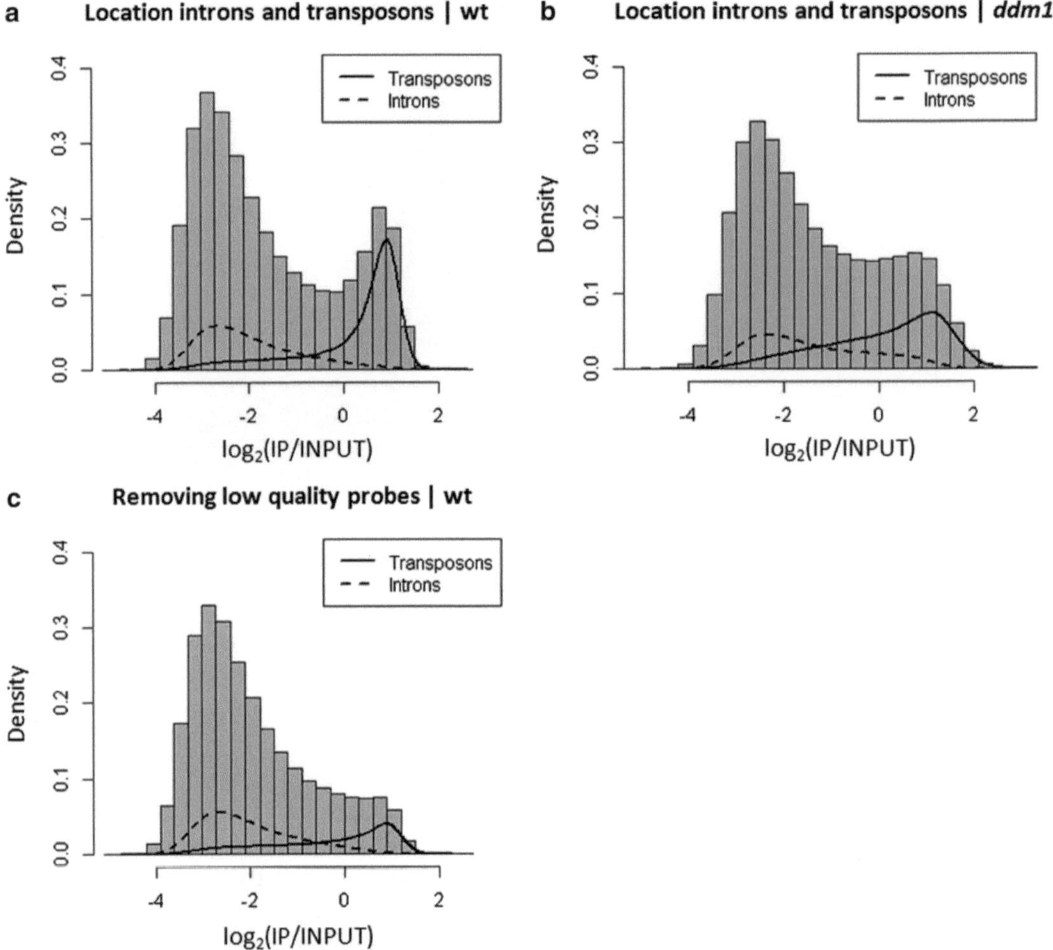


**Fig. 5** Probe signal distributions of transposable elements and introns. (**a**) The $\log_2$(IP/INPUT) signal distribution of one dye combination (IP: *green*, INPUT: *red*) of the wild type Columbia plant with transposons (*solid*) and introns (*dashed*) highlighted. (**b**) Same as in (**a**) but shown for a *ddm1* mutant plant which has lost 70 % of its DNA methylation. The intron distribution is not much affected by this loss. (**c**) Same as in (**a**) but with low quality probe signals removed. As can be seen, the intron distribution is robust to low quality probe signals

```
3 CHR01FS000007129
4 CHR01FS000007479
5 CHR01FS000007814
6 CHR01FS000008139
```

For plotting purposes and further analysis steps, it is also necessary to know the rows of the probes that correspond to transposons or introns. The following commands determine those rows:

```
> rows_intron    <- which(wt_green_red[,1] %in%
                    p_id_intron[,1])
> rows_transp    <- which(wt_green_red[,1] %in%
                    p_id_transp[,1])
```

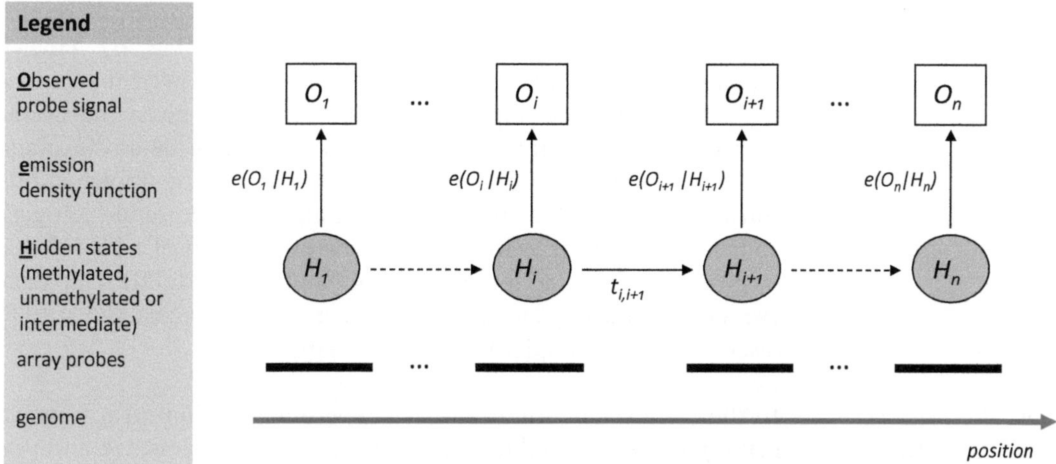


**Fig. 6** Schematic of a HMM model in the context of genome-wide tiling array data. An explanation of the different components of the HMM is provided in the figure

```
> rows_intron_highq    <- setdiff(rows_intron,lowq_
                       rows)    #Without flagged
> rows_transp_highq    <- setdiff(rows_transp,lowq_
                       rows)    #probes
```

The file with plotting code contains the code for plotting Fig. 5.

As can be seen in this figure, even within these two extreme annotation sets (i.e., transposons and introns) there is substantial signal variation (Fig. 5). This is probably due to some level of biological variation (i.e., not all transposable element sequences are methylated and not all introns are unmethylated), but it certainly also reflects the limitations of the measurement technology itself [6]. In addition, many probe signals belong to annotation sets that cannot be easily assigned to these extreme mixture components, and their classification as methylated, unmethylated, or intermediate is inherently probabilistic.

Our principle analytical approach for performing this probabilistic classification relies on the use of a HMM. A Markov chain is a list of random values $\{H_1, H_2, \ldots, H_n\}$ that satisfy the so-called Markov property: the value at position $i$ ($H_i$) is related solely to the values at positions $i-1$ and $i+1$ ($H_i-1$ and $H_i+1$), with given transition probabilities. In the case of a Hidden Markov chain, an output $\{O_1, O_2, \ldots, O_n\}$ is observed that depends on the unobserved (hidden) states of the chain, $\{H_1, H_2, \ldots, H_n\}$ [8]. In the case under consideration, the output or observed chain is the $\log_2$ transformed IP/INPUT signal ratio, while the hidden chain is the methylation state of the DNA sequence indexed by the array probe (Fig. 6).

Hence, the HMM approach capitalizes on two key properties of MeDIP-chip data: (1) probe signals are noisy proxies of an unobserved (hidden) methylated, intermediate, or unmethylated state, and (2)

the probe signals are spatially correlated along the genome, so that neighboring probes provide similar information (Fig. 6). HMMs account for these two properties and provide a powerful statistical framework for classifying individual probe signals given the overall data structure. Our implementation goal is to provide a robust and fast model estimation procedure. We achieve this by implementing software code in C++ and by incorporating several useful biological constraints. In what follows, we outline our version of a HMM that is specifically designed for *Arabidopsis* NimbleGen MeDIP-chip data. We start by detailing key data preparation steps before we move on to discuss the actual implementation strategy.

#### *3.5.1 Data Rescaling Using Intron Probes*

In the context of a single MeDIP experiment within-array normalization is not required in our experience. Nonetheless, we find that rescaling the overall signal distribution is generally a good idea to permit more meaningful comparisons across different individuals (i.e., experimental conditions), should such additional data become available. To achieve this, we make use of the intron probe signal distribution (Fig. 7a). We standardize this distribution and express the overall signal distribution in terms of their standard deviation values. This has the effect of placing the mean of the intron probe signal at zero and rescaling the values as standard deviation values. This rescaling can be implemented with the following code:

```
> intron_mean   <- mean(wt_dye_swap[rows_intron,2])
> intron_sd     <- sd(wt_dye_swap[rows_intron,2])
> wt_dye_swap_rs <-     (wt_dye_swap[,2]-intron_mean)/
                  intron_sd
```

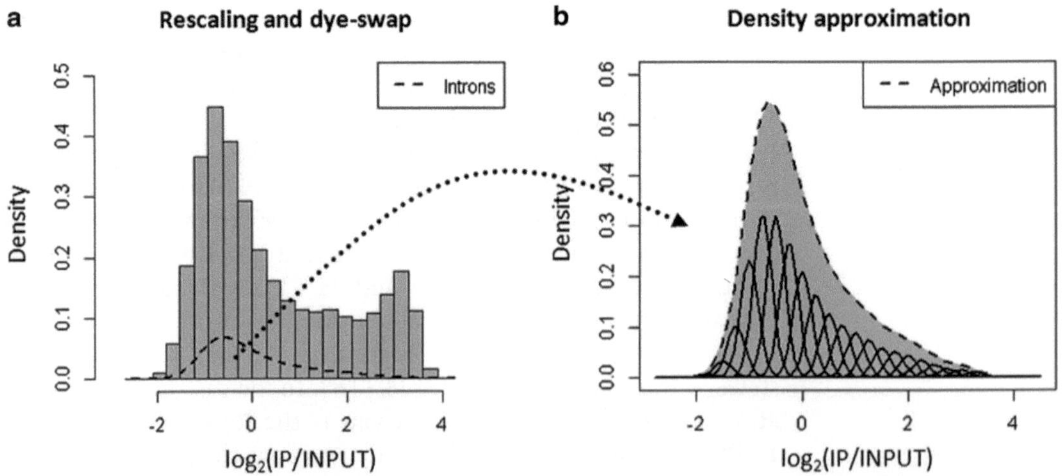


**Fig. 7** Data rescaling and density approximation probe signal distribution of introns. (**a**) Original signal distribution with intron density highlighted (*dashed line*). (**b**) Density of the signal distribution for introns approximated using a mixture of a large number of Gaussian distributions with fixed variance and equally spaced means

```
> wt_dye_swap_rs          <- data.frame(wt_dye_swap[,1],
                             wt_dye_swap_rs)
> names(wt_dye_swap_rs)   <- c("PROBE_ID","DYE_SWAP_RS")
```

One can use the plotting code that is provided as a text file to plot the density of the rescaled data (Fig. 7a).

We find that the intron signal distribution can be safely used for this rescaling process, insofar that it is relatively invariant to high levels of experimental variation. To illustrate this in the context of an extreme case, we compare the signal distribution for wild type to that for the *ddm1* mutant, in which DNA methylation is reduced approximately 70 %. The MeDIP-chip experiment reflects this methylation loss nicely (Fig. 5b), with the signal distribution being clearly reduced in height over the enriched component. Clearly, the signal distribution for intronic sequences is not noticeably affected in *ddm1*, as expected.

### 3.5.2 Implementation of the Hidden Markov Model

We apply our HMM to the rescaled data following a two-step process. First, we use the Baum–Welch algorithm [8, 9] to estimate the best model parameters given the observed probe signals (Fig. 6). Second, we find the most likely hidden sequence of probe states given these estimated parameters. A copy of the C++ code that we implement ccan be found at the above URL (*see* end of Subheading 2.3).

A characteristic feature of our HMM implementation is the use of biologically meaningful constraints on the emission probability density functions, $e(O_i \mid H_i)$, during the Baum–Welch estimation procedure (Fig. 6). In the following we outline these assumptions. A summary of them can be found in Table 5. Alternatively, all the parameters of the emission probabilities could be freely estimated by means of the Baum–Welch algorithm, but we find that a more biologically meaningful approach is preferable.

*Emission probability of unmethylated hidden state*: We employ the signal distribution for introns to obtain an approximation of

**Table 5**
**Summary of the constraints for the emission probability density functions used in the Baum–Welch algorithm**

| Hidden state | Distribution | Parameters |
|---|---|---|
| Unmethylated states | Intron signal distribution | Estimated as a mixture of 30 normals (EM algorithm) with fixed variance |
| Methylated states | Gaussian | Mean: fixed at the 99th quantile of the intron signal distribution. Variance: freely estimated |
| Intermediate states | Gaussian | Mean: fixed at ½ (mean of the methylated distribution). Variance: equal to the variance of the methylated distribution |

the emission probability of the unmethylated hidden state (Fig. 7a). In this way we incorporate biological knowledge of introns being mostly unmethylated directly into the estimation procedure. This bypasses the need to explicitly assume an emission density function, and also speeds up computation. We approximate the signal distribution for introns to an arbitrary degree using mixtures of a large number of Gaussian random variables (Fig. 7b). Estimation is carried with the EM algorithm [10], which can be implemented using the following code:

```
> density_approx    <- function(data,mu,var,lambda,eps,
                       num_norm){
+   mu_diff  <- mu[2]-mu[1]
+   min_val  <- mu[1]-5*mu_diff
+   max_val  <- mu[num_norm]+5*mu_diff
+   rows_extr        <- which(data < min_val | data >
                       max_val)
+   if(length(rows_extr) > 0){   # Remove extreme
+    data    <- data[-rows_extr] # data points
+   }
+   loglik_diff      <- 100000     # Initial loglik diff
+   counter        <- 0       # Iteration counter
+   dnorm_tot        <- rep(0,length(data))
+   for(A in 1:num_norm){
+     dnorm_tot      <-
+     dnorm_tot + lambda[A]*dnorm(data,mean=mu[A],sd
      =sqrt(var))
+   }
+   loglik_pre       <- sum(log(dnorm_tot))    #
    Initial loglik
+   while(loglik_diff > eps){   # Estimate mixture
+     counter        <- counter+1
+     for(A in 1:num_norm){      # Update lambda
+       post         <-          # Posterior prob
+       lambda[A]*dnorm(data,mean=mu[A],sd=sqrt(
        var))/dnorm_tot
+       lambda_new   <- sum(post)/length(data)
+       lambda[A]    <- lambda_new
+     }
+     dnorm_tot      <- rep(0,length(data))
+     for(A in 1:num_norm){
+       dnorm_tot    <-
+       dnorm_tot + lambda[A]*dnorm(data,mean=mu[A],
        sd=sqrt(var))
+     }
+     loglik_new     <- sum(log(dnorm_tot))          #
      New loglik
+     loglik_diff     <- abs(loglik_new - loglik_pre)
      # New loglik diff
```

```
+     loglik_pre     <- loglik_new
+     cat("Iteration  =  ",counter," Log-lik  diff  =
      ",loglik_diff,"\n")
+    }
+    output   <- list(mu,var,lambda)       # Return results
+    names(output) <- c("mu","var","lambda")
+    return(output)
+ }
>
> mus <- round(seq(-2.75,4.5,0.25),2)
> intron_data <- wt_dye_swap_rs[rows_intron,2]
> den_appr    <- density_approx(data=intron_data,mu=mus,
               var=0.03,
+
lambda=rep((1/30),30),eps=0.1,num_norm=30)
Iteration =  1  Log-lik diff =  64314.34
Iteration =  2  Log-lik diff =  510.154
Iteration =  3  Log-lik diff =  41.84451
Iteration =  4  Log-lik diff =  6.50513
Iteration =  5  Log-lik diff =  1.766472
Iteration =  6  Log-lik diff =  0.7541914
Iteration =  7  Log-lik diff =  0.421424
Iteration =  8  Log-lik diff =  0.2723324
Iteration =  9  Log-lik diff =  0.1922282
Iteration =  10  Log-lik diff =  0.1442565
Iteration =  11  Log-lik diff =  0.1132907
Iteration =  12  Log-lik diff =  0.09209521
>
```

The code for plotting the result (Fig. 7b) is provided as a text file.

We generally find that a fit with 30 Gaussians with fixed variance provides a sufficient approximation (Fig. 7b). Parameter estimates are outputted to be used as input in the Baum–Welch algorithm.

*Emission probability of methylated hidden state*: The second constraint relates to the emission probability for the methylated hidden state. We assume this distribution to be Gaussian, with mean fixed to the 99th quantile of the emission probability of the unmethylated state (i.e., the signal distribution for introns). The variance of the distribution is estimated freely by the Baum–Welch algorithm.

*Emission probability of intermediate hidden state*: The last constraint relates to the emission probability for the intermediate hidden state. We assume again this distribution to be Gaussian, with a mean that is fixed between the mean of the emission probability of the unmethylated hidden state (i.e., the intron distribution) and the mean of the emission probability of the methylated hidden state. We take the variance of this distribution to be equal to the variance of the emission probability of the methylated hidden state.

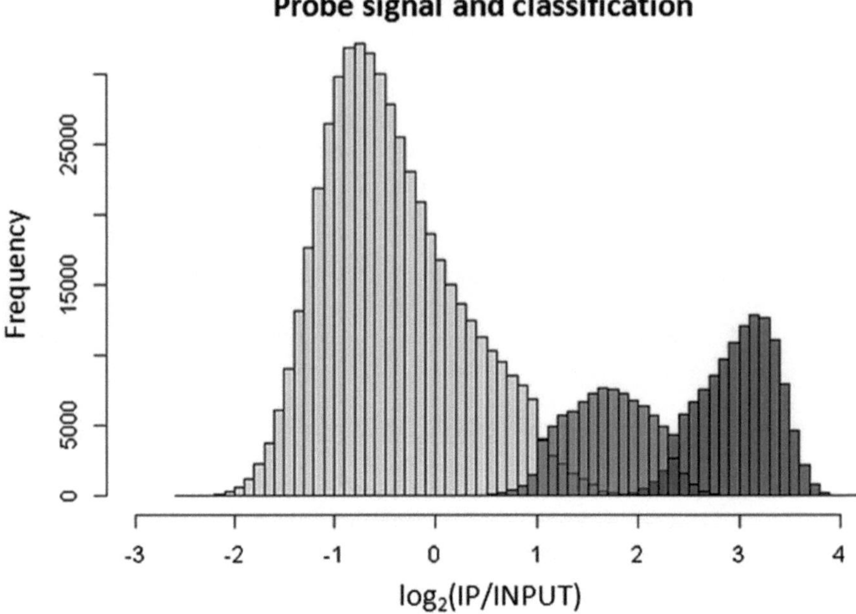


**Fig. 8** The $\log_2$(IP/INPUT) signal distribution for a wild type *A. thaliana* Col-0 accession. Probes are classified into unmethylated probes (*light gray*), intermediate probes (*gray*), and methylated probes (*dark gray*)

The following code generates files that are used as input for the Hidden Markov program written in C++:

```
> values <- c(den_appr$mu,den_appr$var,den_
        appr$lambda)
> parameters <- c(paste("mu",1:30,sep=""),"var_all",
+ paste("lambda",1:30,sep=""))
> para_est <- data.frame(parameters,values)
> names(para_est) <- c("PARAMETER","VALUE")
> write.table(para_est,"para_wild_type.txt",quote=FALSE,
  sep="\t",
+ row.names=FALSE,col.names=TRUE)
> write.table(wt_dye_swap_rs,"dye_swap_signal_wild_
  type.txt",
+ quote=FALSE,sep="\t",row.names=FALSE,col.
  names=TRUE)
```

Once all the free parameters of the HMM have been estimated, we proceed to infer the most likely hidden sequence of probe states given the parameters of the HMM and the observed probe signals. There are several possible strategies, depending on our optimality criterion. We consider two cases: (1) finding the single best hidden sequence of probe states, given the observed probe signals and the parameters of the HMM (the so-called Viterbi algorithm) (Viterbi 1967, Rabiner 1989), or (2) finding the single hidden probe state which is individually most likely at each position, given the observed probe signals and the parameters of the HMM (Rabiner 1989). A copy of the C++ code implemented for the

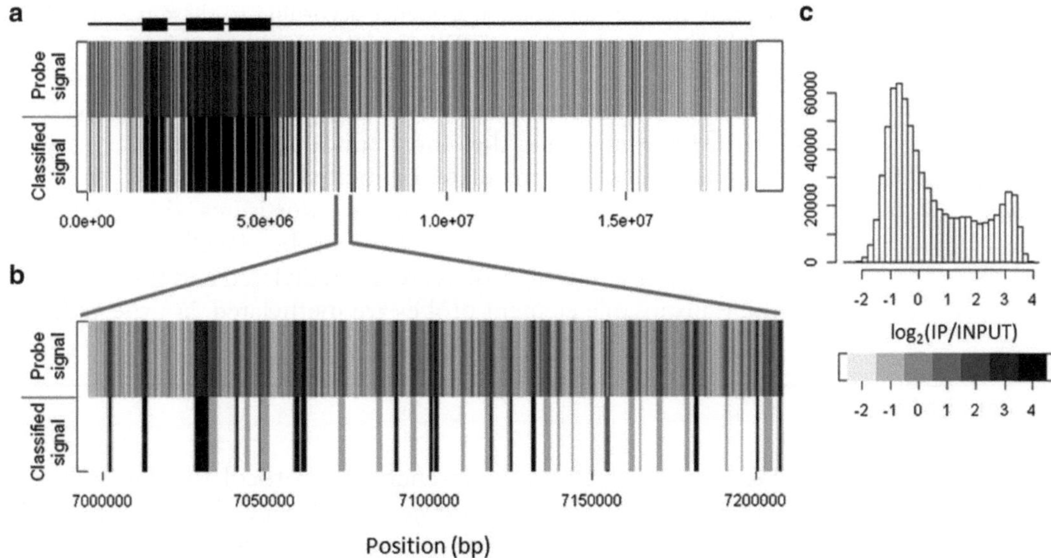


**Fig. 9** Example probe classification. (**a**) The probe signal (*top*) and the corresponding (*hidden*) DNA methylation state (*bottom*) of chromosome 4 in wild type Columbia accession plotted against position (base pairs, *x*-axis); methylated (*black*), unmethylated (*white*), and intermediately methylated (*gray*). As expected, we find substantial methylation in the pericentromeric regions as well as in the heterochromatic knob present on the short arm of the chromosome. (**b**) Magnification of a small region on chromosome 4: we can see how the $\log_2$(IP/INPUT) signal of each probe (*top*) is assigned to methylated, intermediate, or unmethylated state, depending on its signal and on the signal of its surrounding probes. (**c**) Color code for the probe signal density plot, with the corresponding probe density distribution

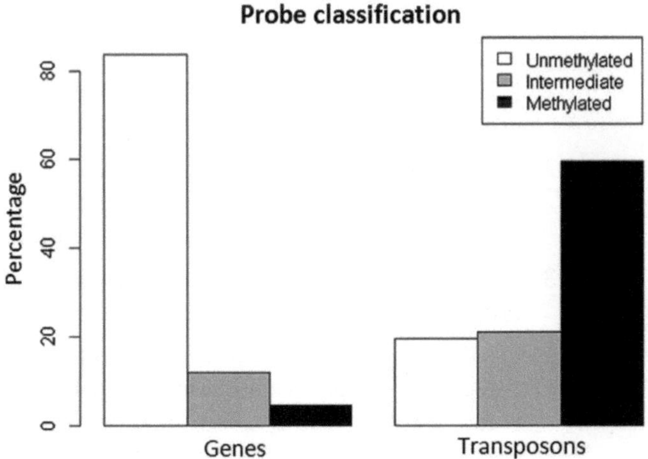


**Fig. 10** Probe classification of genes and transposable elements

identification of the optimal sequence according to these two definitions can be found at the above URL (*see* end of Subheading 2.3).

### 3.6 Graphical and Biological Assessment of HMM Results

The above algorithms probabilistically classify the original log2(IP/INPUT) signals to the three underlying methylation states (unmethylated, intermediate, or methylated) (Fig. 8). This "hidden chain" of methylation states constitutes the methylome (Fig. 9). Annotation analysis of the probe classification (Fig. 10) shows that most gene probes are unmethylated and the majority of the transposable element probes are methylated, as expected.

## 4 Conclusions

We have described a comprehensive protocol for the analysis of DNA methylomes in *Arabidopsis* using MeDIP tiling arrays. Our protocol uniquely combines all necessary steps from "wet lab" to "dry lab" to begin to characterize the epigenetic landscape in this species. Owing to the relatively favorable cost of tiling array technology over more recent deep sequencing approaches, our protocol can be easily scaled up to population-level studies. Such large epigenetically informative approaches will soon become an indispensable tool in the context of intra- or intergenerational functional studies [12]. We have applied the protocol outlined here to a large panel of epiRILs [3] in order to characterize the role of DNA methylation in complex trait inheritance.

## 5 Notes

1. For more than 250 μL of beads, separate in two tubes for washes.
2. Transfer to new tubes decreases noise. This is done in classical tubes because siliconized tubes tend to leak too much with phenol/chloroform and can cause loss of material.
3. MinElute cleaning is a critical step as the efficiency of WGA2 drops dramatically without it.
4. PicoGreen quantification is very sensitive. Be careful to homogenize your samples well before quantification. Since PicoGreen is not stable in light, quantification must be done soon (less than 30 min) after addition of PicoGreen and samples should be maintained in the dark before use.
5. It is important to verify incorporation of dye using the following formula: concentration in DNA (pmol/μL)/concentration in Dye (pmol/μL). Values are usually between 100 and 180.

## Acknowledgements

This work was supported in part by grants from the Agence Nationale de la Recherche (Genoplante TAG project, to V.C.) and by the European Union Network of Excellence "The Epigenome" (to V.C.). S.C. was supported by a Ph.D. studentship from the Ministère de l'Enseignement Supérieur et de la Recherche. R.W., M.C.-T., and F.J. were supported by grants from The Netherlands Organisation for Scientific Research.

### References

1. Johannes F, Porcher E, Teixeira FK, Saliba-Colombani V, Simon M, Agier N, Bulski A, Albuisson J, Heredia F, Audigier P, Bouchez D, Dillmann C, Guerche P, Hospital F, Colot V (2009) Assessing the impact of transgenerational epigenetic variation on complex traits. PLoS Genet 5:e1000530
2. Reinders J, Wulff BB, Mirouze M, Marí-Ordóñez A, Dapp M, Rozhon W, Bucher E, Theiler G, Paszkowski J (2009) Compromised stability of DNA methylation and transposon immobilization in mosaic *Arabidopsis* epigenomes. Genes Dev 23:939–950
3. Colomé-Tatché M, Cortijo S, Wardenaar R, Morgado L, Lahouze B, Etcheverry M, Martin A, Feng S, Duvernois-Berthet E, Labadie K, Wincker P, Jacobsen SE, Jansen RC, Colot V, Johannes F (2012) Features of the Arabidopsis recombination landscape resulting from the combined loss of sequence variation and DNA methylation. Proc Natl Acad Sci USA 109:16240–16245.
4. Cokus SJ, Feng S, Zhang X, Chen Z, Merriman B, Haudenschild CD, Pradhan S, Nelson SF, Pellegrini M, Jacobsen SE (2008) Shotgun bisulphite sequencing of the *Arabidopsis* genome reveals DNA methylation patterning. Nature 452:215–219
5. Lister R, O'Malley RC, Tonti-Filippini J, Gregory BD, Berry CC, Millar AH, Ecker JR (2008) Highly integrated single-base resolution maps of the epigenome in *Arabidopsis.* Cell 133:523–536
6. Johannes F, Wardenaar R, Colomé-Tatché M, Mousson F, de Graaf P, Mokry M, Guryev V, Timmers HT, Cuppen E, Jansen RC (2010) Comparing genome-wide chromatin profiles using ChIP-chip or ChIP-seq. Bioinformatics 26:1000–1006
7. Laird PW (2010) Principles and challenges of genome-wide DNA methylation analysis. Nat Rev Genet 11:191–203
8. Rabiner LR (1989) A tutorial on Hidden Markov Models and selected applications in speech recognition. Proc IEEE 77: 257–286
9. Baum LE, Petrie T, Soules G, Weiss N (1970) A maximization technique occurring in the statistical analysis of probabilistic functions of Markov chains. Ann Math Stat 41: 164–171
10. McLachlan GJ, Peel D (2000) Finite mixture models. John Wiley and Sons, Inc.
11. Viterbi AJ (1967) Error bounds for convolutional codes and an asymptotically optimum decoding algorithm. IEEE Trans Inf Theory 13:260–269
12. Johannes F, Colot V, Jansen RC (2008) Epigenome dynamics: a quantitative genetics perspective. Nat Rev Genet 9:883–890

# Chapter 10

# Methylation-Sensitive Amplified Polymorphism (MSAP) Marker to Investigate Drought-Stress Response in Montepulciano and Sangiovese Grape Cultivars

**Emidio Albertini and Gianpiero Marconi**

## Abstract

Methylation-sensitive amplified polymorphism (MSAP) is a technique developed for assessing the extent and pattern of cytosine methylation and has been applied to genomes of several species (Arabidopsis, grape, maize, tomato, and pepper). The technique relies on the use of isoschizomers that differ in their sensitivity to methylation.

**Key words** DNA methylation, Methylation-sensitive amplified polymorphism, MSAP, Denaturing PAA gel, Silver staining, Sequencing, Drought-stress, *Vitis vinifera*

## 1 Introduction

Methylation of DNA in plants has been associated with regulation of gene expression, genome defense, cell differentiation, chromatin inactivation, and genomic imprinting (*see* Preface to this volume). Higher plants are predominately methylated at the dinucleotide CpG and trinucleotide CpXpG in the form of 5-methylcytosine. In particular, in higher plants, cytosine methylation is distributed between the sequences 5mCG and 5mCNG [1], mainly CTG, CAG [2, 3], and CCG [4]. In higher plants, heavy cytosine methylation has also been found to play an important role in gene expression [5–7] as promoter regions of silent genes have been found to be more methylated than actively transcribed sequences [8, 9]. Moreover, DNA methylation in plants is generally species, tissue, organelle, and age specific. In fact, changes in DNA methylation are present throughout the entire life cycle of plants, starting from seed germination up to the plant death either programmed or induced by various agents such as biotic and abiotic stresses. Significant differences in levels of cytosine methylation have been observed among various tissue types in rice [10],

Charles Spillane and Peter C. McKeown (eds.), *Plant Epigenetics and Epigenomics: Methods and Protocols*, Methods in Molecular Biology, vol. 1112, DOI 10.1007/978-1-62703-773-0_10, 

tomato [11], and maize [12] and can be explained as part of the regulation of gene expression during development and differentiation [8, 13, 14]. How methylation regulates gene expression is still not clear [15].

Several methods have been developed for detecting DNA methylation and, recently, the methylation-sensitive amplified polymorphism (MSAP) technique was developed to assess the extent and pattern of cytosine methylation in the genomes of several species [15–19]. The technique is based on the use of the isoschizomers *Hpa*II and *Msp*I that differ in their sensitivity to methylation of their recognition sequences (for further details, *see* the description provided in the Chapter 14). Both enzymes recognize the tetranucleotide sequence 5′-CCGG-3′, but their action is affected by the methylation state of the external or internal cytosine residues. *Hpa*II is inactive when either or both of the two cytosine is fully methylated (both strands methylated) but cleaves the hemimethylated sequence (only one strand methylated), whereas *Msp*I cleaves hemi or fully methylated C5mCGG but not 5mCCGG [20].

Here, we report the example of the application of the MSAP technique to *Vitis vinifera* with the aim of looking for genes related to drought stress. The interest in the identification of drought-resistant genotypes of *V. vinifera* that can optimize their water use is dramatically increasing, especially in areas where it is difficult to extend the irrigation or which are undergoing a progressive shift towards subtropicalization. Previous studies demonstrated that in the presence of severe, multiple summer stresses, Sangiovese cultivar showed morphobiochemical and physiological behaviors that result in the irreversible photoinhibition and partial death of plant leaves, whereas Montepulciano cultivar does not. In our experiments, Montepulciano (tolerant) and Sangiovese (susceptible) cultivars were compared using the MSAP with the aim of understanding if drought-stress tolerance involves DNA methylation. Sequence information revealed that stress-related methylation events affected genes involved in photosynthesis and respiration in plants. These results suggest that the MSAP technique is very useful for isolating genes, which are under epigenetic control.

## 2 Materials

### 2.1 DNA Extraction

1. Preheat a water bath or heating block to 65 °C.
2. Checking concentrate Buffers AP1 and AP3/E of the Qiagen DNeasy Plant Mini Kit (Qiagen) for presence of any precipitates; if necessary dissolve them a 65 °C for 5′.
3. Add ethanol to Buffers AP3/E and AW.
4. Add RNase A to Buffer AP1 before use only.

5. Add the Buffer AP1 with 1 % (w/v) PVP40 (Polyvinylpyrrolidone, Sigma-Aldrich).
6. Perform all centrifugation steps at room temperature (15–25 °C).

### 2.2 Methylation-Sensitive Amplified Polymorphisms

1. Genomic template DNA: 350 ng (concentration should be ~12 ng/μL) in a final volume of 30 μL (*see* **Note 1**).
2. 5× RL buffer: 5× buffer 4 (NEB), 25 mM DTT, and 0.25 μg BSA (*see* **Note 2**).
3. *Eco*RI adapter and *Msp/Hpa* adapter (see below for preparation of adapters): *Eco*RI_A1 (5′-CTCGTAGACTGCGTACC-3′), *Eco*RI_A2 (5′-AATTGGTACGCAGTC-3′), *Msp/Hpa*_A1 (5′-GACGATGAGTCTAGAA-3′), and *Msp/Hpa*_A2 (5′-CGTTCTAGACTCATC-3′).
4. *Eco*RI restriction enzyme (20 U/μL).
5. *Msp*I restriction enzyme (20 U/μL).
6. *Hpa*II restriction enzyme (10 U/μL).
7. 10 mM ATP.
8. T4 DNA ligase (5 U/μL).
9. 1:10 diluted aliquot of digested and ligated template DNA.
10. 10× PCR buffer (Invitrogen) usually supplied by the manufacturer of the enzyme. It may or may not contain magnesium chloride.
11. 50 mM $MgCl_2$.
12. dNTP stock: 2 mM each of dATP, dCTP, dGTP, and dTTP.
13. *Taq* DNA polymerase (5 U/μL) (Invitrogen).
14. PCR primer *Eco*RI+1: 50 ng/μL (5′-GACTGCGTACCAA TTCN-3′, where N means one of the four nucleotides selected by the scientist).
15. PCR primer *Msp/Hpa*+1: 50 ng/μL (5′-GATGAGTCTAGAA CGGN-3′, where N means one of the four nucleotides selected by the scientist).
16. 1:10 diluted aliquot of preamplified DNA.
17. PCR primer *Eco*RI+3: 50 ng/μL (5′-GACTGCGTACCAATT CNNN-3′) (*see* **Note 3**).
18. PCR primer *Msp/Hpa*I+3: 50 ng/μL (5′-GATGAGTCTAGAA CGGNNN-3′) (*see* **Note 3**).
19. Double-distilled sterile water ($ddH_2O$).

### 2.3 Denaturing Polyacrylamide (PAA) Gel

1. Bind-silane and Repel-silane (GE Healthcare) (*see* **Note 4**).
2. Gel stock solution: 8 M urea, 6 % acrylamide: *N′*,*N′*-methylene bisacrylamide (19:1 ratio) in 1× TBE. Dissolve 288 g urea,

34.2 g acrylamide, and 1.8 g bisacrylamide in $H_2O$, add water to 500 mL, dissolve completely by stirring overnight. Add 30 mL of 20× TBE stock solution and fill up to 600 mL. Pass through filter paper and store in a flask wrapped in tin foil. The solution is stable at room temperature for several weeks.

3. 10 % Ammonium persulfate solution (APS) (freshly prepared; or stored at 4 °C for less than 1 week).
4. TEMED (*N,N,N′,N′*-Tetramethylethylenediamine, Sigma).
5. 1, 0.5, and 10× TBE buffer (same stock solution can be used for gel and running buffer) pH 8.0.
6. Loading buffer: 98 % formamide, 0.2 % Dextran blue, and 10 mM EDTA. Store in aliquots at −20 °C.

### 2.4 Silver Staining

1. Fix solution: 200 mL acetic acid + 1,800 mL Milli-Q water.
2. Staining solution: 3 g $AgNO_3$ +2 L $H_2O$ +3 mL 37 % formaldehyde.
3. Developing solution: 60 g $NaCO_3$ + 3 mL 37 % formaldehyde + 400 mL sodium thiosulfate (10 mg/mL).
4. Double-distilled water and Milli-Q water.

### 2.5 Amplification of Excised Bands from PAA

1. Lancet.
2. 10× PCR buffer (Invitrogen) usually supplied by the manufacturer of the enzyme. It may or may not contain magnesium chloride.
3. 50 mM $MgCl_2$.
4. dNTP stock: 2 mM each of dATP, dCTP, dGTP, and dTTP.
5. *Taq* DNA polymerase (5 U/μL) (Invitrogen).
6. PCR primer *Eco*RI+3: 50 ng/μL (5′-GACTGCGTACCAA TTCNNN-3′) and PCR primer *Msp/Hpa*+3: 50 ng/μL (5′-GATGAGTCTAGAACGGNNN-3′) used in selective MSAP amplification.
7. Double-distilled water.

### 2.6 PCR Purification and Sequencing

1. Exo-SAP-IT (GE Healthcare).
2. BigDye® Terminator v3.1 (Applied Biosystems).
3. BigDye® XTerminator™ Purification Kit (Applied Biosystems).

## 3 Methods

### 3.1 DNA Extraction

1. Collect young leaves from tolerant (Montepulciano) and susceptible (Sangiovese) plants, freeze in liquid nitrogen and then store at −80 °C.

2. Grind leaves under liquid nitrogen using mortar and pestle.
3. Use the resulting powder (about 100 mg) for isolating DNA with the Qiagen DNeasy Plant Mini Kit (Qiagen) and following the manufacturer's instruction with slight modifications: add 1 % of PVP40 (Polyvinylpyrrolidone, Sigma-Aldrich) to the AP1 extraction buffer, add 180 μL of AP2 buffer to each sample instead of 130 μL, and centrifuge microtubes for 10 min instead of 5 min after ice incubation.
4. Elute DNA in 200 μL of ultra-pure water.
5. Dilute an aliquot of the DNA sample in distilled water (usually in a ratio 1:100; e.g., 5 μL/500 μL) in a microcuvette. Determine the optical density (OD) at 260 and 280 nm against a blank (water). Calculate the DNA concentration in the samples using the formula 1.0 $OD_{260}$ = 50 μg/mL (under standard conditions, i.e., a 1-cm light path). Pure DNA preparations show an $OD_{260}$ to $OD_{280}$ ratio between 1.8 and 2.0.
6. To check DNA quality, load 50 ng of each samples in a 1 % agarose gel. Include at least one lane per comb of a quantitative molecular weight marker to check both DNA quality and real concentration of quantified samples.

### 3.2 Detection of Methylation-Sensitive Amplified Polymorphisms

1. Prepare a double-stranded adapter specific for *Eco*RI sites: dilute lyophilized *Eco*RI_A1 and *Eco*RI_A2 complementary adapters to 200 pmol/μL (Stock solutions) and add 3 μL of each adapter to 120 μL sterile $H_2O$ (the final *Eco*RI adapter concentration is 5 pmol/μL [5 μM]). Heat the adapter mix at 95 °C for 5 min and then allow the temperature to cool to room temperature.
2. Prepare a double-stranded adapter specific for *Msp/Hpa* sites: dilute lyophilized *Msp/Hpa*_A1 and *Msp/Hpa*_A2 complementary adapters to 200 pmol/μL (stock solutions) and mix 30 μL of both adapters with 120 μL of sterile $H_2O$ (the final *Msp/Hpa* adapter concentration will therefore be 50 pmol/μL [50 μM]). Heat at 95 °C for 5 min, let the solution slowly cool down to room temperature.
3. Perform two separate restriction/ligation reactions. For each genomic DNA sample (for example DNA isolated from a tolerant genotype), perform a restriction/ligation using *Eco*RI/*Msp*I and a restriction/ligation using *Eco*RI/*Hpa*II enzyme combination, respectively. The restriction/ligation mix is described in Table 1. Incubate samples for 4 h at 37 °C, and then heat the reaction for 15 min at 70 °C to inactivate the enzymes. *Digested/ligated DNA can be stored at −20°C for extended periods of time.*
4. Dilute an aliquot of the digested and ligated template DNA 1:10 with double-distilled sterile water.

**Table 1**
**Preparation of restriction/ligation mix**

| Reagent | Working dilution | μL | Final concentration |
|---|---|---|---|
| RL buffer | 5× | 10.00 | 1× |
| *Msp*I-*Hpa*II | 20–10 U/μL | 0.25–0.50 | 5 U |
| *Eco*RI | 20 U/μL | 0.25 | 5 U |
| *Msp*I-*Hpa*II adaptor | 50 pmol/μL | 1.00 | 50 pmol |
| *Eco*RI adaptor | 5 pmol/μL | 1.00 | 5 pmol |
| ATP | 10 mM | 1.00 | 0.2 mM |
| T4 DNA ligase | 5 U/μL | 0.20 | 1 U |
| $H_2O$ | – | 6.30–6.05 | – |
| Genomic DNA | – | 30 | 350 ng |
| Final volume | – | 50.0 | – |

**Table 2**
**Preparation of preamplification mix**

| Reagent | Working dilution | μL | Final concentration |
|---|---|---|---|
| *Eco*RI +1 primer | 50 ng/μL | 1.5 | 75 ng |
| *Msp*/*Hpa* +1 primer | 50 ng/μL | 1.5 | 75 ng |
| dNTPs | 5 mM | 2.0 | 0.2 mM |
| PCR buffer | 10× | 5.0 | 1× |
| $MgCl_2$ | 50 mM | 1.5 | 1.5 mM |
| *Taq* DNA polymerase | 5 U/μL | 0.2 | 1 U |
| $H_2O$ | – | 33.3 | – |
| Diluted RL reaction | – | 5.0 | – |
| Final volume | – | 50.0 | – |

5. Mix the components in a microcentrifuge tube according to Table 2 to perform the preamplification reaction.
6. Perform preamplification PCR using the cycling conditions described in Table 3.
7. Run 10 μL of the resulting preamplification PCR products on a 1 % agarose gel. A homogenous, light, smear should be visible for all.

**Table 3**
**Preamplification cycling conditions**

| Cycle name | Temp (°C) | Time | No. of cycles |
|---|---|---|---|
| Pre-PCR_1 | 72.0 | 1′ | 1 |
| Pre-PCR_2 | 94.0<br>65.0<br>72.0 | 45″<br>30″<br>1′ | 1 |
| *Touch-down PCR | 94.0<br>64.3<br>72.0 | 30″<br>30″<br>1′ | 12 |
| PCR | 94.0<br>55.9<br>72.0 | 30″<br>30″<br>1′ | 20 |
| Elongation | 72.0 | 30′ | 1 |

*Annealing temperature for the 12 touch-down profile cycles: 64.3, 63.6, 62.9, 62.2, 61.5, 60.8, 60.1, 59.4, 58.7, 58.0, 57.3, 56.6.

**Table 4**
**Selective PCR amplification mix**

| Reagent | Working dilution | μL | Final concentration |
|---|---|---|---|
| *Eco*RI +3 primer | 50 ng/μL | 0.60 | 30 ng |
| *Msp/Hpa* +3 primer | 50 ng/μL | 0.60 | 30 ng |
| dNTPs | 5 mM | 0.80 | 0.2 mM |
| PCR buffer | 10× | 2.00 | 1× |
| $MgCl_2$ | 50 mM | 0.80 | 2 mM |
| *Taq* DNA polymerase | 5 U/μL | 0.08 | 0.4 U |
| $H_2O$ | – | 10.12 | – |
| Diluted preamp reaction | – | 5.00 | – |
| Final volume | – | 20.0 | – |

8. Dilute the preamplification products 1:10 in double-distilled sterile water. Store at −20 °C or use immediately for selective amplification.
9. Mix the components for selective amplification reaction in a microcentrifuge tube according to Table 4.
10. Perform selective PCR amplification with the same program as listed in Table 2 except for an initial heating at 72 °C for 1 min (*see* **Note 5**).

### 3.3 Denaturing Polyacrylamide Gel and Silver Staining

1. Glass plate preparation. The two plates are treated with different chemicals and they must not be cross-contaminated. Clean both plates with a detergent (Alconox), then rinse them with deionized water and wash with 95 % ethanol. Treat one plate with Bind Silane and the other one with Repel Silane.
2. Clamp the two thoroughly cleaned glass plates in the electrophoresis apparatus with spacers in between, and seal the spacers with, e.g., 2.5 % agarose, PAA sealing gel, tape, or sealing strips.
3. Pour 80 mL of gel stock solution into a suitable flask, and deaerate under vacuum for 10 min. A vacuum-proof sidearm flask can be used, or a hypodermic needle that is inserted through the rubber top of a small laboratory bottle. The acrylamide stock solution should not be more than 2–4 weeks old.
4. Add 80 μL of TEMED and 300 μL of 10 % APS, swirl gently. Do not allow bubbles to form.
5. Pour the solution between the plates, tilting the gel mould and pouring the gel from one side to avoid air bubbles. Put the slot former in place. If a shark's-tooth comb is used, it should be inserted with the smooth side facing the gel.
6. Polymerization takes approximately 45–60 min (*see* **Note 6**). After this, insert the glass plates with the gel into the electrophoresis apparatus (*see* **Note 7**).
7. Fill the electrophoresis tanks with 1× TBE, remove the slot former, and clean the wells thoroughly using a pipette to remove any unpolymerized acrylamide and urea that could be contaminating the loading area. Pre-run gel for 30 min at 40–55 W (corresponding to ~800–2,200 V and 25–35 mA). A stable temperature of 55 °C should be maintained (*see* **Note 8**).
8. Denature samples (2 μL) by adding 2 μL of loading buffer, and heating for 3 min at 95 °C; after denaturation, immediately transfer samples to water–ice (*see* **Note 9**) and wait for at least 2 min before loading.
9. When the samples are ready for loading, switch off the current. If a shark's-tooth comb is used, reinsert the comb with the teeth facing the gel. Teeth should enter the gel to a depth of about 1–2 mm. Make sure that a sufficiently large gap is left for loading the samples. If appropriate, indicate the sample numbers below the respective wells with a pen. Urea readily diffuses from the gel into the wells, which will prevent the samples from sinking to the bottom. Therefore, the wells may have to be cleaned again before sample application.
10. Use special thin pipette tips or a Hamilton syringe to deposit the sample (2.5–3.5 μL) to the bottom of the well (*see* **Note 10**). It is very useful to load a DNA molecular weight marker to one

or more lanes to determine the approximate molecular weight of interestingly bands.

11. Reconnect electrodes and run the gel for 3–4 h at 40–55 W at 55 °C; these conditions are sufficient to visualize MSAP pattern.
12. When run is complete, allow plates to cool before separating them. Carefully separate the glass plates by lifting the upper plate at one edge using a spatula. The gel is expected to adhere to the plate treated with Bind-silane.
13. Transfer the glass plate that retains the gel to a container, cover with Fixative solution and agitate for 20 min (**Note 11**).
14. Transfer the gel to a fresh container and wash three times with ultrapure water.
15. Incubate the gel in the Staining solution for 30 min.
16. Wash with Milli-Q water.
17. Incubate the gel in the Developer solution. Keep the gel agitated for 6–7 min until the bands become visible.
18. Transfer the gel in the Fix solution for 3 min to terminate the developing reaction.
19. Rinse the gel twice with water.
20. The gel can now be photographed and any interesting bands excised from PAA gel. An example of the MSAP pattern is shown in Fig. 1.

### 3.4 Amplification of Bands Excised from PAA

1. Using a lancet, excise the selected methylated DNA-induced band from PAA gel. An example of such a band is shown in Fig. 2.
2. Place the excised band in a 1.5-mL tube containing 100 μL of double-distilled water for 6 h at 4 °C and stir it frequently (*see* **Note 12**).
3. Centrifuge for 10 min at 13,000 × *g*.
4. Use 5 μL of the supernatant containing resuspended DNA to amplify the band in presence of the same *Eco*RI+3 and *Msp/Hpa*+3 primer combination that yielded the polymorphic band.
5. The reaction mix for the amplification of eluted bands is listed in Table 5.
6. Perform PCR using the following cycling conditions: one cycle of 94 °C for 3 min, 30 cycles of 94 °C for 1 min, 55 °C for 1 min, and 72 °C for 1 min, followed by a final extension step at 72 °C for 10 min.
7. Run 10 μL of PCR product on 1.5 % agarose gel, load at least one lane with a quantitative molecular weight marker to check DNA amplification quality, expected size and quantity (ng/μL) of obtained amplicon.

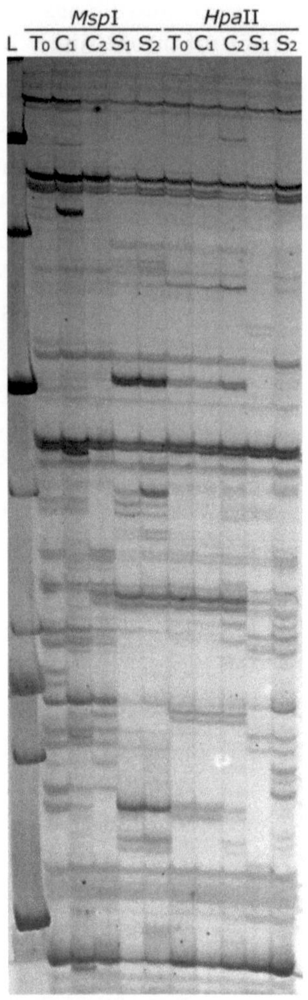


**Fig. 1** Example of an MSAP silver stained PAA gel. One day after the collection of the leaves samples ($T_0$), clones were separated and either regularly irrigated (C samples) or water-stressed (S samples). Leaves were then collected at day 7 ($C_1$ and $S_1$) and day 14 ($C_2$ and $S_2$). DNA was isolated from collected leaves and used for MSAP analysis as described, employing *Msp*I and *Hpa*II in association with the *Eco*RI enzyme

### 3.5 Sequencing Analysis

1. Use the remaining part of PCR product for the ExoSAP reaction. Mix PCR product reaction (about 15 μL) with 1 μL of ExoSAP-IT enzyme mix.
2. Perform reaction at 37 °C for 30 min and then inactivate ExoSAP-IT enzyme mix by heating the mixture at 80 °C for 15 min (*see* **Note 13**). The DNA is now ready for direct sequencing.
3. For sequencing reaction, add the reagents described in Table 6.

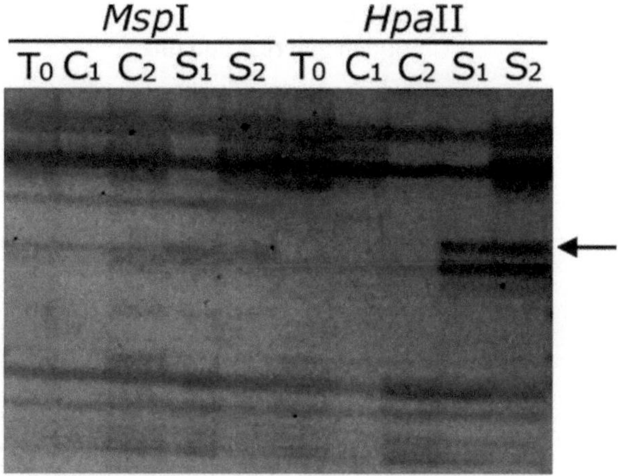


**Fig. 2** Identification of methylation-sensitive polymorphisms by MSAP. Several types of polymorphisms can be visualized by MSAP. In particular, our attention was focused on two classes. In some cases, stress conditions caused methylation of sites and lead to the disappearance of an amplification product; in other cases, stress conditions caused demethylation of sites, which lead to the appearance of an amplification product in the stressed material (indicated by the *arrow*). Lanes are labeled as in Fig. 1

**Table 5**
**PCR mix for the amplification of eluted bands**

| Reagent | Working dilution | μL | Final concentration |
|---|---|---|---|
| *Eco*RI +3 primer | 50 ng/μL | 1.0 | 2 ng |
| *Msp*/*Hpa* +3 primer | 50 ng/μL | 1.0 | 2 ng |
| dNTPs | 5 mM | 1.0 | 0.2 mM |
| PCR Buffer | 10× | 2.5 | 1× |
| $MgCl_2$ | 50 mM | 0.8 | 1.6 mM |
| *Taq* DNA polymerase | 5 U/μL | 0.2 | 1 U |
| $H_2O$ | – | 13.5 | – |
| Resuspended DNA | – | 5.0 | – |
| Final volume | – | 25.0 | – |

4. Perform sequencing reaction using the following cycling conditions: one cycle of 96 °C for 1 min, 30 cycles of 96 °C for 10 s, 50 °C for 5 s, and 60 °C for 4 min.
5. For purification of sequencing products, use BigDye® XTerminator™ Purification Kit. Add 45 μL of SAM™ and 10 μL BigDye® XTerminator™ solution to each sample.

**Table 6**
**Sequencing reaction mix**

| Reagent | Concentration | μL |
|---|---|---|
| Ready reaction premix | 2.5× | 0.5 |
| BigDye seq. buffer | 5× | 2.0 |
| Sequencing primer | 3.2 pmol | 1.0 |
| DNA template | 2 ng/100 bp | – |
| $H_2O$ | – | to 10 |
| Final volume | – | 10.0 |

6. Seal the plate with heat seal film or caps.
7. Thoroughly mix the contents of the plate for 30 min (vortex every 5 min and allow the BigDye® XTerminator™ to precipitate).
8. Centrifuge the plate at 3,000 × *g* for 3 min.
9. Run purified reactions on a DNA Sequencer using BigDye® XTerminator™ run module.
10. Use sequences obtained for each sample as query in searches performed on the National Center for Biotechnology Information (NCBI; http://www.ncbi.nlm.nih.gov) database with BLASTN (est_others database) and BLASTX (nr database) applications [21–23] with the aim of comparing nucleotides and deduced protein against sequenced genomes (*see* **Note 14**).

## 4 Notes

1. DNA preparations need to be of reasonable quality to ensure complete digestion by restriction enzymes.
2. 5× RL buffer can be stored at −20 °C for an extended period of time.
3. The innermost of the three selective bases must be the same as used in the preamplification reaction.
4. After each run, one or both glass plates needs to be siliconized. Treatment with Bind-silane (store dark and dry at room temperature) reagent ensures that the gel adheres to the glass plate after the end of the run. The other plate is treated with Repel-silane (store dark and dry at room temperature). For this treatment, follow the manufacturers' instructions.
5. Twenty cycles for the PCR step in the selective amplification may not be sufficient for species with very large genomes.

6. The gel can be stored at 4 °C overnight in humid condition (e.g., wrapped in tissues moistened with water).
7. Different types of PAA gel apparatus are commercially available; we recommend that the user follow the instructions of the manufacturer.
8. Do not exceed 60 °C because urea may decompose.
9. Water–ice mix allows a perfect contact with tubes used for denaturing samples.
10. Generally, the same tip can be used for loading subsequent samples if rinsed in 1× TBE present in the upper buffer chamber.
11. In the Silver Staining protocol, it is crucial to respect the times for each step. Leaving the gel in each reagent for more or less time could result in failure of the staining procedure.
12. After stirring, check that excised gel band remains soaked in the double-distilled water.
13. It is convenient to do these steps in a thermal cycler.
14. The best choice is perform BLASTN and BLASTX analysis against the nearest genome to your own species.

## References


1. Gruenbaum Y, Navey-Many T, Cedar H, Razin A (1981) Sequence specificity of methylation in higher plant DNA. Nature 292:860–862
2. Pradhan S, Adams RLP (1995) Distinct CG and CNG DNA methyltransferases in *Pisum sativum*. Plant J 7:471–481
3. Kovarik A, Matyasek R, Leitch A, Gazdova B, Fulnecek J, Bezdek M (1997) Variability in CpNpG methylation in higher plant genomes. Gene 204:25–33
4. Jeddeloh JA, Richards EJ (1996) (m)CCG methylation in angiosperms. Plant J 9: 579–586
5. Meyer P, Niedenhof I, Ten Lohuis M (1994) Evidence for cytosine methylation of non-symmetrical sequences in transgenic *Petunia hybrida*. EMBO J 13:2084–2088
6. Ulian EC, Magill JM, Magill CW, Smith RH (1996) DNA methylation and expression of NPT II in transgenic petunias and progeny. Theor Appl Genet 92:976–981
7. Rossi V, Motto M, Pellegrini L (1997) Analysis of the methylation pattern of the maize Opaque-2 (O2) promoter and in vitro binding studies indicate that the O2 B-Zip protein and other endosperm factors can bind to methylated target sequences. J Biol Chem 272: 13758–13765
8. Finnegan EJ, Brettell RIS, Dennis ES (1993) The role of DNA methylation in the regulation of plant gene expression. In: Jost JP, Saluz HP (eds) DNA methylation: molecular biology and biological significance. Birkhauser, Basel
9. Pikaard CS (1999) Nucleolar dominance and silencing of transcription. Trends Plant Sci 4: 478–483
10. Dhar MS, Pethe VV, Gupta VS, Ranjekar PK (1990) Predominance and tissue specificity of adenine methylation in rice. Theor Appl Genet 80:402–408
11. Messeguer R, Ganal MW, Steffens JC, Tanksley SD (1991) Characterization of the level, target sites and inheritance of cytosine methylation in tomato nuclear DNA. Plant Mol Biol 16: 753–770
12. Lund G, Messing J, Viotti A (1995) Endosperm-specific demethylation and activation of specific alleles of alpha-tubulin genes of *Zea mays* L. Mol Gen Genet 246:716–722
13. Finnegan EJ, Peacock WJ, Dennis ES (2000) DNA methylation, a key regulator of plant development and other processes. Curr Opin Genet Dev 10:217–223
14. Richards EJ (1997) DNA methylation and plant development. Trends Genet 13:319–323
15. Xiong LZ, Xu CG, Saghai Maroof MA, Zhang Q (1999) Patterns of cytosine methylation in an elite rice hybrid and its parental lines, detected by a methylation-sensitive amplification polymorphism technique. Mol Gen Genet 261:439–446

16. Ashikawa I (2001) Surveying CpG methylation at 5′-CCGG in the genomes of rice cultivars. Plant Mol Biol 45:31–39
17. Liu B, Brubaker CL, Mergeai G, Cronn RC, Wendel JF (2001) Polyploid formation in cotton is not accompanied by rapid genomic changes. Genome 44:321–330
18. Xu M, Li X, Korban SS (2000) AFLP-based detection of DNA methylation. Plant Mol Biol Rep 18:361–368
19. Portis E, Acquadro A, Comino C, Lanteri S (2004) Analysis of DNA methylation during germination of pepper (*Capsicum annuum* L.) seeds using methylation-sensitive amplification polymorphism (MSAP). Plant Sci 166: 169–178
20. McClelland M, Nelson M, Raschke E (1994) Effect of site-specific modification on restriction endonucleases and DNA modification methyltransferases. Nucleic Acids Res 22: 3640–3659
21. Altschul SF, Lipman DJ (1990) Protein database searches for multiple alignments. Proc Natl Acad Sci U S A 87:5509–5513
22. Altschul SF, Gish W, Miller W, Myers EW, Lipman DJ (1990) Basic Local Alignment Search Tool. J Mol Biol 215:403–410
23. Altschul SF, Madden TL, Schaffer AA, Zhang JH, Zhang Z, Miller W, Lipman DJ (1997) Gapped BLAST and PSI-BLAST: a new generation of protein database search programs. Nucleic Acids Res 25:3389–3402

Chapter 11

# Detecting Histone Modifications in Plants

Jie Song, Bas Rutjens, and Caroline Dean


## Abstract

Histone modifications play an essential role in chromatin-associated processes including gene regulation and epigenetic inheritance. It is therefore very important to quantitatively analyze histone modifications at both the single gene and whole genome level. Here, we describe a robust chromatin immunoprecipitation (ChIP) method for *Arabidopsis*, which could be adapted for other plant species. This method is compatible with multiple downstream applications including qPCR, tilling arrays, and high-throughput sequencing.


**Key words** Histone modification, Epigenetics, Chromatin immunoprecipitation (ChIP)

## 1 Introduction

In eukaryotes, DNA is wrapped around a histone octamer consisting of two copies of H2A, H2B, H3, and H4. Histone tails can be covalently modified at various amino acids (e.g., H3 lysine 4/9/27/36) and be of varying forms (e.g., mono/di/trimethylation and acetylation) [1, 2]. The modifications can be detected by immunoprecipitation (IP) with specific antibodies against the modification.

There are basically two types of chromation IP (ChIP), distinguished by different methods of chromatin conservation and DNA fragmentation prior to the IP. The most commonly used is X-ChIP, where the chromatin is cross-linked and then the DNA sheared by sonication. The other is native ChIP, where nuclei are extracted in their native form without cross-linking, and the DNA fragmented by nuclease digestion. We focus here on the X-ChIP procedure as it is a robust and nonbiased technique favored for analysis of histone modifications.

Several chromatin IP protocols have been described for *Arabidopsis* [3–5] and other plant species, e.g., maize [6] and tomato [7]. Here, we describe in detail a method used extensively in our work on *FLC* epigenetic silencing. It has been streamlined by incorporating new products, for example use of Chelex-100

Charles Spillane and Peter C. McKeown (eds.), *Plant Epigenetics and Epigenomics: Methods and Protocols*, Methods in Molecular Biology, vol. 1112, DOI 10.1007/978-1-62703-773-0_11, © Springer Science+Business Media New York 2014

resin instead of high salt reverse cross-linking [8] and optimized for use with a large number of samples where quantitative data is required.

## 2 Materials

Stock solutions (*see* Subheading 2.1) are prepared and stored at room temperature. Buffers used in the experiment are prepared fresh on the day and kept at 4 °C, unless otherwise stated. β-mercaptoethanol (TOXIC, use in fume-hood) and protease inhibitors are added to the solution just prior to use.

### 2.1 Stock Solutions

1. Phosphate-buffered saline (PBS) buffer (10×): NaCl 1.3 M, $Na_2HPO_4$ 30 mM, $NaH_2PO_4$, pH 7.4.
2. Sucrose, 2 M.
3. Tris–HCl, 1 M pH 8.
4. $MgCl_2$, 1 M.
5. Triton X-100, 10 % (w:v).
6. EDTA, 0.5 M pH 8.
7. NaCl, 5 M.
8. SDS, 10 % (w:v).
9. LiCl, 4 M.
10. NP-40, 10 % (w:v).

### 2.2 Chromatin Extraction and DNA Fragmentation

1. Cross-linking solution: 1 % (v:v) formaldehyde (Sigma) in PBS buffer, freshly prepared at room temperature.
2. Quenching solution: 2 M glycine, freshly dissolved in $ddH_2O$ at room temperature.
3. Extraction buffer 1: 0.4 M sucrose, 10 mM Tris–HCl, pH 8.0, 5 mM β-mercaptoethanol, and protease inhibitor cocktail (cOmplete, Roche).
4. Extraction buffer 2: 0.25 M sucrose, 10 mM Tris–HCl, pH 8.0, 10 mM $MgCl_2$, 1 % (v:v) Triton X-100, 5 mM β-mercaptoethanol, and protease inhibitor cocktail (cOmplete, Roche).
5. Extraction buffer 3: 1.7 M sucrose, 10 mM Tris–HCl, pH 8.0, 0.15 % (v:v) Triton X-100, 2 mM $MgCl_2$, 5 mM β-mercaptoethanol, and protease inhibitor cocktail (cOmplete, Roche).
6. Nuclei lysis buffer: 50 mM Tris–HCl, pH 8.0, 10 mM EDTA, 1 % (w:v) SDS, and protease inhibitor cocktail (cOmplete, Roche).

### 2.3 Immunoprecipitation

1. ChIP dilution buffer: 1.1 % (v:v) Triton X-100, 1.2 mM EDTA, 16.7 mM Tris–HCl, pH 8.0, 167 mM NaCl, and protease inhibitor cocktail (cOmplete, Roche).
2. Antibodies against specific histone modifications [9] and no antibody control. We use rabbit polyclonal antibody anti-H3K27me3 (Millipore, 07-449) to detect H3K27me3, and rabbit polyclonal antibody anti-H3 (Abcam, ab1791) to detect H3 (*see* **Note 1**). For these rabbit polyclonal antibodies, rabbit IgG (Millipore) is used as the mock control.
3. Dynabeads protein A (Invitrogen) and magnet (*see* **Note 2**).
4. Low salt wash buffer: 150 mM NaCl, 0.1 % (w:v) SDS, 1 % (v:v) Triton X-100, 2 mM EDTA, and 20 mM Tris–HCl, pH 8.0.
5. High salt wash buffer: 500 mM NaCl, 0.1 % (w:v) SDS, 1 % (v:v) Triton X-100, 2 mM EDTA, and 20 mM Tris–HCl, pH 8.0.
6. LiCl wash buffer: 0.25 M LiCl, 1 % (v:v) NP-40, 1 % (w:v) sodium deoxycholate, 1 mM EDTA, and 10 mM Tris–HCl, pH 8.0.
7. TE buffer: 10 mM Tris–HCl, pH 8.0 and 1 mM EDTA.
8. Elution: 10 % (w:v) Chelex 100 resin (Bio-Rad).
9. Proteinase K solution: 20 mg/mL Proteinase K (Roche) in water, stored at −20 °C.

### 2.4 DNA Clean-Up

1. Phenol/chloroform solution, mixture of phenol, chloroform, isoamyl alcohol in the ratio of 25:24:1 (v:v:v) (Sigma).
2. NaAc, 3 M pH 5.2.
3. GlycoBlue (Ambion).
4. Ethanol, absolute and 75 % (v:v).

### 2.5 Quantitative PCR

SYBR Green qPCR mix (Roche) and LightCycler 480 II Instrument (Roche), or an alternative quantification system.

## 3 Methods

The work flow for ChIP is illustrated in Fig. 1. The whole procedure can be performed within 2 days, Day 1 for formaldehyde cross-linking, chromatin extraction, DNA fragmentation, and immunoprecipitation, and Day 2 for immunoprecipitation washes and quantification. There are possible pausing points, for example after formaldehyde cross-linking and after DNA fragmentation, as mentioned in respective steps below.

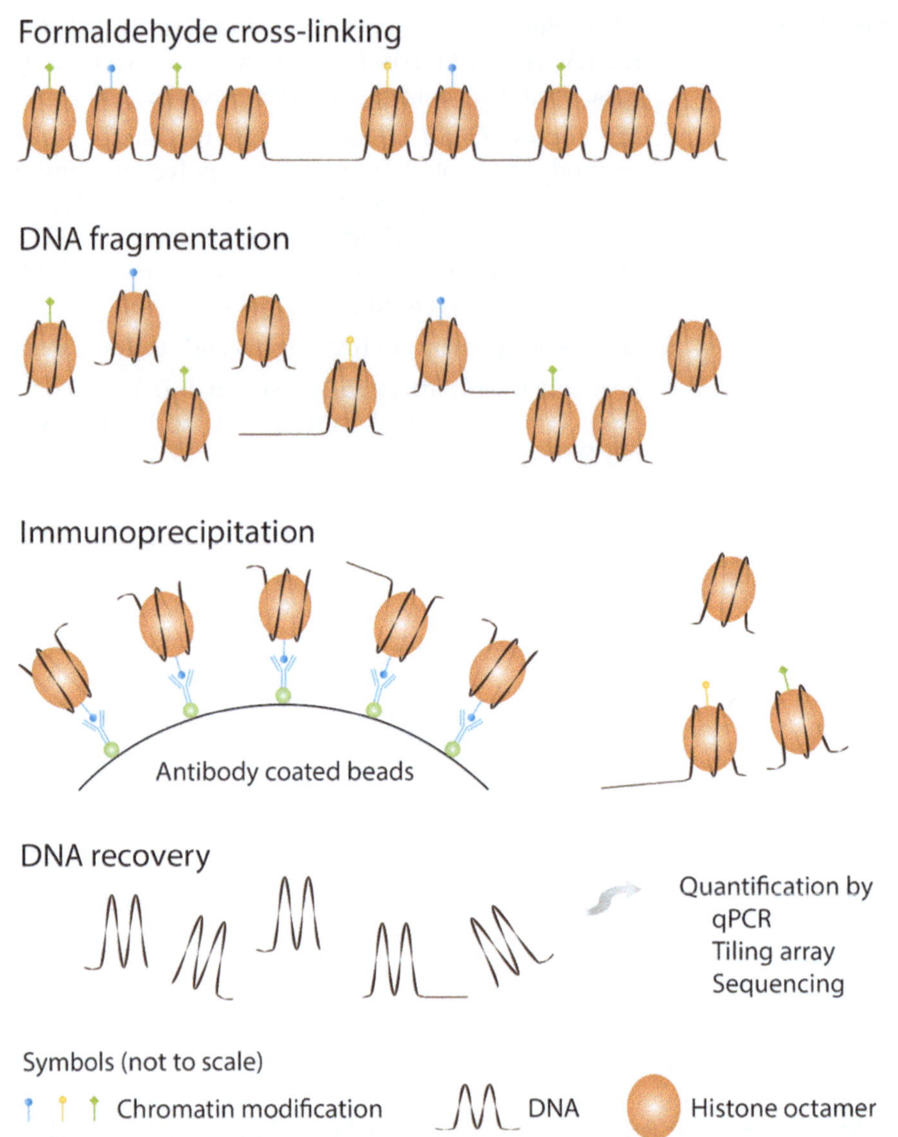


**Fig. 1** Work flow of chromatin immunoprecipitation (ChIP)

### *3.1 Plant Growth and Formaldehyde Cross-Linking*

1. Plant materials are grown on media or compost under desirable conditions. If grown on nutrient media, we recommend Murashige and Skoog (MS) minus glucose in order to limit bacterial growth.
2. Harvest 1–2 g plant materials in a 50-mL Falcon tube (*see* **Note 3**).
3. Rinse plant material twice in dd$H_2O$ and remove water throughout after the second rinse.
4. Submerge plant materials in 37 mL of cross-linking solution. Stuff the tube with nylon mesh to keep plants immersed in the buffer. Vacuum infiltrate at room temperature for three times,

5 min each time and release vacuum in between to allow buffer penetrating plant tissues. Shake desiccator slightly to remove air bubbles. At this stage, seedlings should appear "water-soaked" or translucent (*see* **Note 4**).

5. Stop the cross-linking by addition of glycine quenching solution to a final concentration of 0.125 M (2.5 mL of 2 M stock into 37 mL of cross-linking buffer) and application of vacuum for additional 5 min.
6. Rinse off formaldehyde with 40 mL dd$H_2O$ for at least three times.
7. Remove water as thoroughly as possible by placing seedlings on a paper towel before freezing in liquid nitrogen. At this stage, cross-linked material can be either stored at −80 °C for up to several months or processed further for chromatin extraction.

### 3.2 Chromatin Extraction and DNA Fragmentation

All steps in this and the following Subheading 3.3 are carried out at 4 °C. Refrigerator centrifuges and motors are precooled to 4 °C.

1. Grind plant materials in liquid nitrogen to a fine powder.
2. Resuspend the powder in 30 mL extraction buffer 1 in a 50-mL Falcon tube. Vortex immediately to mix and incubate for 5 min or until solution is homogenous. Make sure there is the volume of the buffer is more than five times more than that of the powder.
3. Filter the solution through a double layer of Miracloth into a new, ice-cold 50 mL Falcon tube. Repeat until solution is clear.
4. Spin the filtered solution for 20 min at 2,880 × *g* at 4 °C.
5. Gently remove supernatant and resuspend the pellet in 1 mL of extraction buffer 2.
6. Transfer the solution to 1.5-mL Eppendorf tube.
7. Centrifuge at 12,000 × *g* for 10 min at 4 °C.
8. Remove the supernatant and resuspend pellet in 300 μL of extraction buffer 3.
9. Overlay the resuspended pellet onto 900 μL of extraction buffer 3 in a fresh 1.5-mL Eppendorf tube.
10. Spin for 45 min to 1 h at 16,000 × *g*.
11. Remove the supernatant and resuspend the chromatin pellet in a DNA LoBind tube (Eppendorf) with 320 μL of nuclei lysis buffer by pipetting up and down. From this point on, DNA LoBind tubes or similar products are used to reduce sample-to-surface binding.

    *Check point: save a 10 μL aliquot representing "unsheared" chromatin for later examination.*

12. Once resuspended, sonicate the chromatin solution for three times, 5 min each time (30 s on/off intervals) at the "Low" setting using BioRupter (Diagenode) according to the manufacturer's instructions (*see* **Note 5**). In between runs, cool the water bath by adding more ice and correcting the water level if necessary. This fragments DNA into manageable sizes, usually ranging from 200 to 800 bp and centering at 500 bp (*see Check point* below).
13. The sonicated chromatin solution can be frozen in liquid nitrogen and stored at −80 °C or processed further for immunoprecipitation.
14. Spin the sonicated chromatin suspension for 5 min at 16,000 × *g* to pellet debris.

   *Check point: save a 10-μL aliquot representing "sheared" chromatin, together with "unsheared" chromatin taken from* step 11, *to assess sonication efficiency* (*see* **Note 6**, Fig. 2).

### 3.3 Immunoprecipitation

1. Prepare Dynabeads protein A magnetic beads. Use 15 μL beads per IP, one tube for the mock control, one tube for each antibody. Use 4 μg of anti-H3k27me3 antibody per H3K27me3 IP and 3 μg of anti-H3 antibody per H3 IP (*see* **Note 7**).

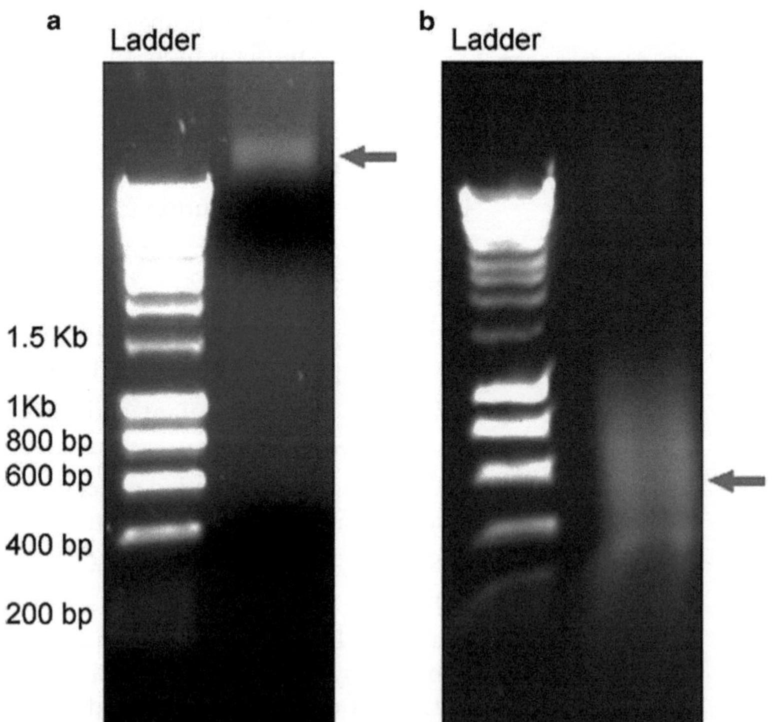


**Fig. 2** DNA fragmentation by sonication. Before (**a**) and after (**b**) sonication with BioRupter (Diagenode) used at a LOW setting for 3 pulses of 5 min each (30 s on/off intervals), laddered using HyperLadder I (Bioline). *Arrows* indicate DNA

2. Wash beads three times in ChIP dilution buffer. Let beads attach to the magnet and discard buffer. Add antibody and incubate in 50 μL ChIP dilution buffer for 1 h (*see* **Note 8**).
3. Wash prepared antibody coated beads three times with ChIP dilution buffer.
4. Take 10 % volume of the chromatin solution (usually 30 μL) as input control. The input control samples need to be processed for DNA recovery as described in Subheading 3.4.
5. Transfer the remaining chromatin solution (270 μL) into a Falcon tube and dilute ten times with ChIP dilution buffer to dilute the 1 % SDS to 0.1 %.
6. Add 900 μL diluted chromatin solution to each tube with antibody coated beads, mock control, H3K27me3 IP and H3 IP, respectively, and incubate rotating overnight or at least 2.5 h.
7. Apply following washes, 1 mL of wash buffer per IP sample (*see* **Note 9**). Attach beads to magnet after each wash to collect beads and discard supernatant.
   (a) Low salt wash buffer, two washes, 5 min each.
   (b) High salt wash buffer, one wash, 5 min.
   (c) LiCl wash buffer, one wash, 5 min.
   (d) TE buffer, two washes, 5 min each.
8. During the last TE wash, transfer the beads to a new LoBind tube to further lower background (optional).
9. After the final TE wash, remove TE buffer thoroughly. Carry out the following step immediately.

### 3.4 DNA Recovery and Clean-Up

1. Elute immune complexes by adding 100 μL 10 % Chelex resin and incubating at 95 °C, 1,300 rpm for 10 min. Input samples, taken from **step 4**, Subheading 3.3, need to be treated the same way from this step onwards to recover DNA (*see* **Note 10**).
2. Cool on ice and collect contents by brief spin. Add 2 μL Proteinase K solution and incubate at 50 °C for 30 min to digest proteins.
3. Boil for another 10 min at 95 °C, 1,300 rpm.
4. Brief spin to collect contents. Add $ddH_2O$ to the sample to make up the volume to 500 μL. Spin and transfer supernatant to a new LoBind tube.
5. Use phenol–chloroform extraction to remove protein (*see* **Note 11**). Precipitate DNA with 1/10 vol NaAc, 2 μL GlycoBlue, and 2 vol absolute ethanol. Spin at top speed to pellet DNA and wash with 75 % ethanol. Pellet DNA again after each wash by centrifugation at top speed for 3 min. Air-dry DNA pellet and resuspend in $ddH_2O$ for quantitative PCR (qPCR). Input samples can be

further diluted 2–10 times to obtain similar concentration of DNA as the IP samples. DNA resuspension volume can be adjusted according to the requirements of subsequent assays.

### 3.5 Quantification and Normalization

The ChIPped DNA is quantified in qPCRs to examine enrichment of target sequences. Primers are designed using Primer3Plus [10] to cover sequences of interests and controls. Raw data from qPCRs are analyzed using the $2^{-[\mathrm{Delta}][\mathrm{Delta}]}C_T$ analysis method [11]. Data are represented as fold enrichment to the mock control (usually used for protein–DNA binding but not histone modifications), percentage of input (*ChIP H3K27me3/Input*)*100 %, or enrichment per nucleosome (*ChIP H3K27me3/ChIP H3*)*100 % (Fig. 3a).

The enrichment of a certain histone modification at a specific locus can be expressed in relation to a reference sequence, an internal control. This normalization is particularly relevant when comparing across different treatments and experiments, in which case variations between experiments need to be taken into account. The normalization can be calculated as:

$$(\text{Target sequence H3K27me3} / \text{control sequence H3K27me3}) * 100\ \%,$$

or

[(Target sequence H3K27me3/H3)/(control sequence H3K 27me3/H3)] * 100 % (Fig. 3b).

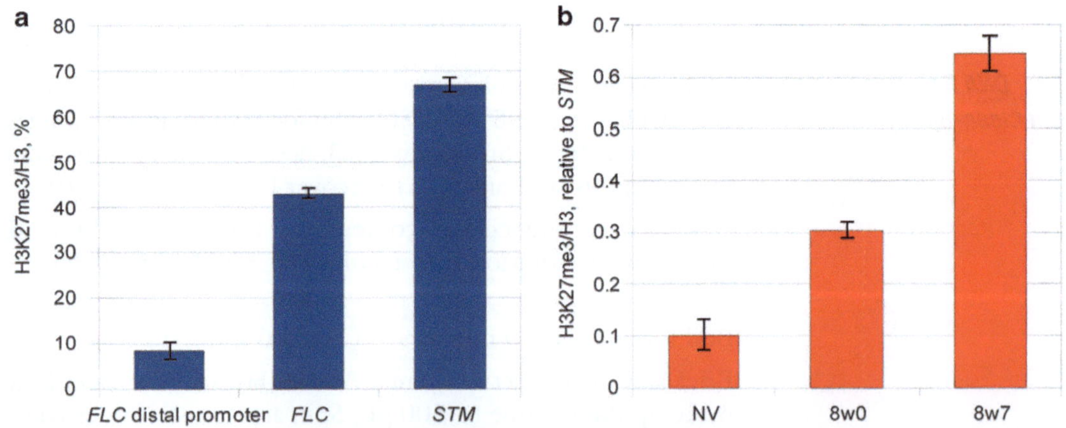


**Fig. 3** Detection of H3K27me3 on *Arabidopsis* seedlings. (**a**) H3K27me3 in vernalized seedlings. Data are presented as (ChIP H3K27me3/ChIP H3)*100 %. *FLC* distal promoter quantified using primers 5′-atccagaaaagggcaaggag-3′ and 5′-cgaatcgattgggtgaatg-3′; *FLC* locus quantified using primers 5′-ctttttcatgggcaggatca-3′ and 5′-tgacatttgatcccacaagc-3′; *SHOOT MERISTEMLESS* (*STM*) locus quantified using 5′-gcccatcatgacatcacatc-3′ and 5′-gggaactactttgttggtggtg-3′. (**b**) H3K27me3/H3 on *FLC* locus (primers 5′-ctttttcatgggcaggatca-3′ and 5′-tgacatttgatcccacaagc-3′), before (nonvernalized, NV), during (at the end of 8 weeks vernalization treatment, 8w0), and after vernalization (7 days post vernalization, 8w7). Data are calculated as [(*FLC* ChIP H3K27me3/H3)/(control sequence STM ChIP H3K27me3/H3)]*100 %

In the case of H3K27me3, as it has been mapped across the *Arabidopsis* genome [12], loci constantly covered with high levels of H3K27me3 can be used as positive internal controls, e.g., *SHOOT MERISTEMLESS* (*STM*), or *AGAMOUS* (*AG*) [13]. Loci with low levels of H3K27me3 can be used as a negative internal control, e.g., *ACTIN* [14].

## 4 Notes

1. There are commercially available, ChIP tested antibodies against histone modifications [9]. New antibodies need to be tested for ChIP with controls. There is no guarantee that antibodies, which work in other applications, for example Western blotting, would definitely work in ChIP.
2. Magnetic beads (for example Dynabeads from Invitrogen) have advantages over agarose beads. Magnetic beads are collected gently by placing in a magnetic field, with no columns or centrifugations involved. There is no bad volume remaining after each wash, which helps reduce background.
3. Mature plant tissue contains a high proportion of vacuole to cytoplasm, extensive secondary thickening and complex cell walls. These properties could significantly reduce efficiency of nuclei extraction so we recommend using young tissue if possible.
4. This is one of the key steps in ChIP. When working with a different type of tissue or plant species, cross-linking should be optimized to best preserve chromatin structure and yet not to make the subsequent reverse cross-linking too difficult. Optimization of cross-linking was described by Haring et al. [6]. In brief, after cross-linking, extract chromatin as described in Subheading 3.2, **steps 1–11**, take an aliquot (e.g., 10 μL) to extract free DNA using phenol–chloroform solution. Take another aliquot and reverse cross-linking as described in Subheading 3.4 to recover DNA. Run both samples on agarose gel to check cross-linking and reverse cross-linking efficiency. Be aware that there will still be free DNA with harsh cross-linking conditions; this "open" chromatin with a loss of nucleosome is typically at genomic loci that are actively transcribed [15].
5. The purpose of sonication is to shear DNA into manageable sizes. This can be achieved using water bath, e.g., BioRupter (Diagenode), or probe-based sonication device. Sample needs to be kept ice-cold, and rested on ice in between pulses. At this stage, chromatin solution can be viscous. Ensure there are no air bubbles in the samples. If using a probe-based sonicator,

the amplitude setting and duration are subjective and may vary between each application.

6. The "sheared" and "unsheared" chromatin samples are processed to recover DNA as described in Subheading 3.4, treat with RNase to remove RNA and run on 1.5 % agarose gel to check sonication efficiency. DNA fragmentation needs to be optimized when first time working on a different biological sample.
7. The amount of antibody used in one assay is usually recommended by its manufacturer, typically 1–10 μg. Concentrations of antibody in each batch may vary.
8. Binding reaction volume can be adjusted according to the manufacturer's instructions. Excessive antibodies are washed away during the subsequent washes with ChIP dilution buffer.
9. In our experience, these wash steps generally work fine for histone ChIP. If concern rises regarding weaker antibody–antigen binding and loss of ChIP signal, some steps, for example the high salt wash, can be skipped or salt concentration can be reduced. In contrast, if background level is high, wash steps can be repeated to reduce nonspecific binding. For histone ChIP, due to its abundance, the signal-to-noise ratio is usually more satisfactory compared to transcription factor ChIPs.
10. Chelex-100 resin is a quick way to reverse cross-linking [8]. Alternatively, it can be achieved by elution of bead-bound complexes using elution buffer (1 % [w:v] SDS, 0.1 M $NaHCO_3$, incubation at 65 °C for 15 min), followed by high salt (addition of NaCl to a final concentration of 0.2 M) reverse cross-linking at 65 °C for at least 4 h.
11. Protein in the sample is digested by proteinase K in previous steps. However, we find that traces of protein left in the sample, especially in input samples, may interfere with ChIP quantification. Protein can be further removed from the sample by phenol–chloroform extraction. In the previous step, $ddH_2O$ is added to the sample to increase the volume so that it is easy to take up the supernatant. An alternative, nontoxic method is to use resin that collects protein, for example StrataClean resin (Stratagene). ChIPped DNA is then precipitated and resuspended in a desirable volume. It can also be purified using commercially available PCR clean-up kits that collect both single-stranded and double-stranded DNA.

## References

1. Margueron R, Reinberg D (2010) Chromatin structure and the inheritance of epigenetic information. Nat Rev Genet 11:285–296
2. Turner BM (2007) Defining an epigenetic code. Nat Cell Biol 9:2–6
3. Lippman Z, May B, Yordan C, Singer T, Martienssen R (2003) Distinct mechanisms determine transposon inheritance and methylation via small interfering RNA and histone modification. PLoS Biol 1:e67
4. Bastow R, Mylne JS, Lister C, Lippman Z, Martienssen RA, Dean C (2004) Vernalization requires epigenetic silencing of FLC by histone methylation. Nature 427:164–167
5. De Lucia F, Crevillen P, Jones AM, Greb T, Dean C (2008) A PHD-Polycomb repressive complex 2 triggers the epigenetic silencing of FLC during vernalization. Proc Natl Acad Sci U S A 105:16831–16836
6. Haring M, Offermann S, Danker T, Horst I, Peterhansel C, Stam M (2007) Chromatin immunoprecipitation: optimization, quantitative analysis and data normalization. Plant Methods 3:e11
7. Ricardi M, Gonzalez R, Iusem N (2010) Protocol: fine-tuning of a chromatin immunoprecipitation (ChIP) protocol in tomato. Plant Methods 6:e11
8. Nelson JD, Denisenko O, Sova P, Bomsztyk K (2006) Fast chromatin immunoprecipitation assay. Nucleic Acids Res 34:e2
9. Egelhofer TA, Minoda A, Klugman S, Lee K, Kolasinska-Zwierz P, Alekseyenko AA, Cheung M-S, Day DS, Gadel S, Gorchakov AA, Gu T, Kharchenko PV, Kuan S, Latorre I, Linder-Basso D, Luu Y, Ngo Q, Perry M, Rechtsteiner A, Riddle NC, Schwartz YB, Shanower GA, Vielle A, Ahringer J, Elgin SCR, Kuroda MI, Pirrotta V, Ren B, Strome S, Park PJ, Karpen GH, Hawkins RD, Lieb JD (2011) An assessment of histone-modification antibody quality. Nat Struct Mol Biol 18: 91–93
10. Rozen S, Skaletsky H (2000) Primer3 on the WWW for general users and for biologist programmers. Methods Mol Biol 132:365–386
11. Livak KJ, Schmittgen TD (2001) Analysis of relative gene expression data using real-time quantitative PCR and the 2-[Delta][Delta]CT method. Methods 25:402–408
12. Zhang X, Clarenz O, Cokus S, Bernatavichute YV, Pellegrini M, Goodrich J, Jacobsen SE (2007) Whole-genome analysis of histone H3 lysine 27 trimethylation in Arabidopsis. PLoS Biol 5:e129
13. Finnegan EJ, Dennis ES (2007) Vernalization-induced trimethylation of histone H3 lysine 27 at FLC is not maintained in mitotically quiescent cells. Curr Biol 17:1978–1983
14. Yu X, Michaels SD (2010) The Arabidopsis Paf1c complex component CDC73 Participates in the modification of FLOWERING LOCUS C chromatin. Plant Physiol 153:1074–1084
15. Giresi PG, Kim J, McDaniell RM, Iyer VR, Lieb JD (2007) FAIRE (formaldehyde-assisted isolation of regulatory elements) isolates active regulatory elements from human chromatin. Genome Res 17:877–885

# Chapter 12

# Quantitatively Profiling Genome-Wide Patterns of Histone Modifications in *Arabidopsis thaliana* Using ChIP-seq

**Chongyuan Luo and Eric Lam**

## Abstract

Genome-wide quantitative profiling of chromatin modifications is a critical experimental approach to study epigenetic and transcriptional control mechanisms. Since first being reported in 2007, chromatin immunoprecipitation followed by high-throughput sequencing (ChIP-seq) has soon became a popular method-of-choice for profiling chromatin modifications and transcription factor-binding sites in eukaryote genomes. ChIP-seq has the advantage over the earlier ChIP-chip approach in multiple aspects including the lower amount of input DNA required, an expanded dynamic range and compatibility with sample multiplexing. Here we describe a detailed protocol for profiling histone modification in the *Arabidopsis thaliana* genome with ChIP-seq using the SOLiD™ 2.0 high-throughput sequencing platform. As read length and sequencing depth are two critical factors determining data quality and cost, we have developed bioinformatics approach to evaluate the effect of read length and sequencing depth on the alignment accuracy and the generated chromatin profile, respectively. Our analyses suggest that 2–3 million high quality sequencing tags with a read length of 35 nucleotides would be sufficient to profile the majority of histone modifications in this popular model plant species.

**Key words** ChIP-seq, Chromatin, Histone modifications, *Arabidopsis*

## 1 Introduction

Chromatin structures consist of a complex array of distinct chemical features that are spatially organized into dynamic molecular patterns across the genome. In recent years, some of these features that are relevant to genome organization and epigenetic regulation have been recognized to be the nonrandom incorporations of histone variants, covalent modifications of DNA and histones as well as the modulation of nucleosome positioning. For each feature that contributes to chromatin structure, the patterns of genomic distributions are intimately linked to the corresponding genome response or output. For example, the histone H3 variants H3.3 is incorporated primarily at actively transcribed regions by its chaperone Hira [1, 2]. Trimethylation at lysine 4 of histone H3 subunits

Charles Spillane and Peter C. McKeown (eds.), *Plant Epigenetics and Epigenomics: Methods and Protocols*,
Methods in Molecular Biology, vol. 1112, DOI 10.1007/978-1-62703-773-0_12, 

(H3K4me3) and various histone acetylations are established surrounding the transcription start sites of active genes in euchromatin and may facilitate transcription initiation and elongation events [3, 4]. Therefore the task of elucidating the functions of chromatin structure variants would be aided by the ability to routinely carry out high-resolution mapping of their genomic locations. What is equally important is the ability to accurately quantify the abundance of chromatin structure variations at a given locus. Important discoveries have thus been made recently by quantitatively comparing the abundance of histone modifications between distinct cell types. Through genome-wide profiling of histone modification in embryonic stem cells (ESC) and differentiated cells, the co-localization of H3K4me3 and H3K27me3 at key developmental factors was found to be a characteristic of ESC and may regulate important development genes [5, 6]. These bivalent states largely resolve into H3K4me3 or H3K27me3 monovalent states when ESC differentiates [5, 6]. Cell type-specific ChIP-seq studies have also been recently achieved in plants for two root cell types using a novel nuclei label-and-capture technique called INTACT [7]. Further extension and application of these types of approaches are likely to provide rapid advances in our database of chromatin feature variations that can be correlated with cellular states.

Chromatin immunoprecipitation (ChIP) followed by microarray hybridization (ChIP-chip) or high-throughput sequencing (ChIP-seq) is a method to interrogate the genome-wide localization of particular types of chromatin-related epitopes. These epitopes are typically specific modifications on histone subunits that are known to be epigenetically relevant, or sequence-specific DNA-binding proteins such as transcription factors. Moreover, ChIP-derived strategies have also been devised to "pull-down" cytosine-methylated (5-mC) DNA by using a 5-mC-specific DNA-binding protein and have been used successfully to display the genome-wide methylation landscape [8]. The major steps of ChIP include cross-linking DNA with bound proteins and the subsequent enrichment of DNA associated with a particular protein by a specific antibody. The utilization of genomic tilling array or high-throughput sequencing enables the global survey of DNA samples produced by ChIP. ChIP-seq has several advantages over the ChIP-chip method especially for large genomes. (1) ChIP-seq generally requires substantially less ChIPed DNA than ChIP-chip. A full flowcell of sequencing with SOLiD™ 2.0 generating 400 million tags needs less than 5 ng of library DNA [9]. For ChIP-chip, ChIPed sample needs to be amplified to generate more than 2 μg of DNA for each array [10]. Therefore with equal amount of ChIPed sample, ChIP-seq requires less cycles for library amplifications that usually lead to superior quantification over ChIP-chip. (2) ChIP-seq possesses greater dynamic range as compared to ChIP-chip. The dynamic range of any microarray platform is limited by background hybridization noises at the lower end and signal saturations at the higher end. In contrary, ChIP-seq can detect input DNA species at any abundance with high confidence that is limited only by the number of good quality sequence reads. (3) ChIP-seq has the special advantage to sequence multiplexed samples in one run through the "barcode" or "index" design, which is impossible for any microarray platform [11]. Together with the ever-decreasing cost with concomitant increase in throughput for sequencing, ChIP-seq can be a more economical and rapid technique than ChIP-chip for obtaining the same amount of data with equivalent or better quality.

In this chapter, we describe the procedure to profile histone modifications in the *Arabidopsis* genome by using the SOLiD™ 2.0 sequencing platform. We aim to develop an economical procedure without relying on commercial kits so that it would be affordable even for generating large-scale multiplexing libraries. As the quality and cost of performing ChIP-seq are highly related to the read length and sequencing depth, we have evaluated the impact of read length and sequencing depth on the quality of ChIP-seq experiment and provided our suggestions for the two parameters.

## 2 Materials

### 2.1 Chromatin Immunoprecipitation Using Plant Tissues

1. Formaldehyde (37 % v/v, Fisher Scientific, Catalogue #F79-500, stabilized with 10–15 % methanol).
2. 20 % Triton X-100 (Fisher Scientific, Catalogue #BP151-500), filtered with 0.45 μm filter and stored at room temperature.
3. β-mercaptoethanol (98 %, Fisher Scientific, Catalogue #BP176-100). *Toxic—refer to manufacturer's safety information.*
4. 0.2 M PMSF (Calbiochem, Catalogue #52332) solution, dissolved in ethanol and stored at −20 °C. *Toxic—refer to manufacturer's safety information.*
5. 2 M glycine (Sigma-Aldrich, Catalogue #G7126) solution, filtered with 0.45 μm filter and stored at room temperature.
6. Miracloth (Calbiochem, Catalogue #475855-1R).
7. Probe sonicator (Branson S150).
8. Percoll (Sigma-Aldrich, Catalogue #P-1644).
9. Nuclei isolation buffer (10 mM HEPES pH 7.6, 1 M sucrose, 5 mM KCl, 5 mM $MgCl_2$, 5 mM EDTA pH 8.0), stored at 4 °C; the buffer is supplemented with 1 % formaldehyde, 0.1 % β-mercaptoethanol, 0.6 % Triton X-100, and 0.4 mM PMSF before use.
10. Nuclei separation solution (10 mM HEPES pH 7.6, 1 M sucrose, 5 mM KCl, 5 mM $MgCl_2$, 5 mM EDTA pH 8.0, 15 % Percoll), filtered with 0.45 μm filter and stored at 4 °C.

11. Nuclei lysis buffer (50 mM Tris–HCl pH 7.5, 1 % SDS, 10 mM EDTA pH 8.0), stored at room temperature to avoid the precipitation of SDS.
12. ChIP dilution buffer (15 mM Tris–HCl pH 7.5, 1 % Triton X-100, 150 mM NaCl, 1 mM EDTA), made with autoclaved or filtered solutions and stored at 4 °C.
13. ChIP elution buffer (1 % SDS, 0.1 % $NaHCO_3$), made fresh on the date of use.
14. TE buffer (10 mM Tris–Cl pH 7.5, 1 mM EDTA pH 8.0).
15. 5 M NaCl.
16. Primary antibodies (from various commercial sources as indicated).
17. Protein A agarose beads (Millipore, Catalogue #16-125). Protein A beads blocked with salmon-sperm DNA cannot be used for ChIP-seq since the associated DNA will interfere with subsequent sequencing.
18. Proteinase K (Invitrogen, Catalogue #25530-049).
19. Glycogen (20 mg/ml, molecular biology grade, Fermentas, Catalogue #R0561).
20. MinElute Reaction Cleanup Kit (Qiagen, Catalogue #28204).

### 2.2 Materials for Constructing SOLiD™ Sequencing Libraries

The methods for preparing high-throughput sequencing libraries have been constantly evolving along with the rapid evolution of DNA sequencing instrumentation. As the method described here for generating SOLiD™ libraries was optimized for SOLiD™ 2.0 system, we expect that certain modifications may be necessary for later versions of the SOLiD™ platform and substantial variations would be expected for the other NextGen platforms such as Illumina and 454 technologies. For the ChIP-seq application, one can purchase reagent kits from manufactures of the sequencer or third-party suppliers that would include most necessary reagents for library constructions. However, these reagent kits are relatively expensive and the cost can become prohibitive for generating a large number of multiplexing libraries. Thus we have developed the protocol using separately purchased sequencing adaptors and reagents, which enables the construction of a larger number of libraries with lower costs in order to enable the parallel analysis of multiple chromatin modifications for a number of samples.

1. Covaris S2 (Covaris).
2. T6 (6 × 32 mm) round bottom glass tube with O ring (Covaris, Catalogue #520029).
3. Snap caps (8 mm) for T6 tubes (Covaris, Catalogue #520030).
4. Extended gel loading tip (Fisher Scientific, Catalogue #02-707-181).

5. End-It™ DNA End-Repair Kit (Epicentre Biotechnologies, Catalogue #ER0720).
6. MinElute Reaction Cleanup Kit (Qiagen, Catalogue #28204).
7. Custom synthesized adaptors for constructing fragment libraries with sequences provided in the SOLiD™ System 2.0 Multiplexing Protocol. We have chosen to synthesize the P1 oligos at 50 nmol scales with the PAGE purified grade and the 16 pairs of barcoded P2 oligos at 25 nmol scales with the desalted grade by using a commercial oligonucleotide supplier (Invitrogen). Adaptors are synthesized as single-strand DNA oligos and need to be hybridized as described in Subheading 3.3.1.
8. 5× Green GoTaq™ Reaction Buffer (Promega, Catalogue #M7911).
9. Fast-Link™ DNA Ligation Kit (Epicentre Biotechnologies, Catalogue #LK11025).
10. 40 % Acrylamide/Bis (19:1) solution (Biorad Laboratories, Catalogue #161-0144).
11. Ammonium persulfate (Sigma-Aldrich Catalogue #A3678).
12. 5× TBE buffer.
13. Mini-PROTEAN Tetra Cell (Biorad Laboratories, Catalogue #165-8001).
14. 50 bp DNA ladder (New England Biolab Inc. Catalogue #N3236S) or similar.
15. Custom synthesized P1 and P2 primers for library amplifications with sequences provided in the SOLiD™ System 2.0 Multiplexing Protocol. Oligos were synthesized at 25 or 50 nmol scale with desalted grade.
16. FailSafe™ PCR System with buffer E (Epicentre Biotechnologies, Catalogue #FS99100).
17. QIAquick Gel Extraction Kit (Qiagen, Catalogue #28704).
18. Nanodrop 1000 Spectrophotometer (Thermo Scientific).

## 3 Methods

### 3.1 Chromatin Immunoprecipitation with Aerial Tissues of *Arabidopsis*

The ChIP procedure described here includes a nuclei isolation step, which is optional for a successful ChIP. We also routinely perform ChIP with total plant lysate cross-linked by vacuum-infiltration with formaldehyde. Both methods generate satisfactory results and we have not noticed any difference between results obtained with the two methods. Although the method was originally designed for leaf tissues, it should be readily applicable to roots, flowers, or siliques (for further detail on the isolation of nuclei from inflorescences, see the chapter of Weinhofer and Köhler elsewhere in this volume).

1. Transfer 25 ml of nuclei isolation buffer into a 50 ml centrifuge tube. Add 700 μl of 37 % (v/v) formaldehyde solution, 750 μl of 20 % Triton X-100, 25 μl of β-mercaptoethanol, and 50 μl of 0.2 M PMSF to the nuclei isolation buffer.
2. Grind 0.5–1.5 g of *Arabidopsis* leaf tissue in liquid nitrogen and immediately transfer the powder into the prepared nuclei isolation buffer. Invert the tube several times to completely resuspend the powder in the buffer so that it will be thawed quickly and evenly.
3. Incubate at room temperature for 10 min to cross-link the plant tissue. Invert the tube several times during the incubation for better mixing.
4. Add 1.7 ml of 2 M glycine and incubate at room temperature for 5 min to quench the cross-linking reactions.
5. Filter the lysate through one layer of Miracloth to remove the debris.
6. Pellet the nuclei by centrifugation at 3,000 × *g* for 10 min at 4 °C. The pellet should be largely white indicating most of the chloroplasts are dissolved by the detergent.
7. Discard the supernatant and suspend the pellet in 300 μl of nuclear isolation buffer.
8. Carefully lay the suspended nuclei pellet on 500 μl of nuclei separation solution and centrifuge at 3,000 × *g* for 5 min at 4 °C with a microfuge. A loose pellet with pale white color should be visible at the bottom of the tube.
9. Discard the supernatant and suspend the pellet in 600 μl of nuclei lysis buffer (*see* **Note 1**). Cool the tube on ice for 10 min before starting the sonication.
10. Sonicate the nuclear lysate seven times with power setting 2 for 10 s each. Incubate the tube on ice for 3–5 min between each round of sonication to avoid overheating and foaming.
11. Pellet the debris by centrifugation at 13,000 × *g* for 3 min with a microfuge. Transfer the cleared supernatant to a new tube and discard the pellet. If not used immediately, chromatin samples can be snap frozen with liquid nitrogen and stored at −80 °C.
12. Dilute the chromatin sample ten times with ChIP dilution buffer and aliquot the sample into six microfuge tubes at 1 ml each. One aliquot will be used for the mock ChIPed sample. Set aside 50 μl of the chromatin sample to serve as the "5 % Input" control sample.

13. Add ~2 μg of antibody and 20–40 μl of Protein A agarose beads into the chromatin sample and rotate at 4 °C for 1 h to overnight for ChIP. The optimal amount of antibody for ChIP can be highly variable depending on the titer of the particular antibody. In our experience, 2 μg is a reasonable starting point for most of antibodies that we have tested.
14. Centrifuge at 100 × *g* for 1 min to gently collect the agarose beads, and discard the supernatant. For all steps that agarose beads need to be pelleted, we perform centrifugation at 100 × *g* for 1 min.
15. Rotate the beads with 500 μl of ChIP dilution buffer for 10 min at 4 °C to wash away nonspecific binding materials. Repeat the washing three times.
16. Wash the beads with 500 μl of TE to remove the residual ChIP dilution buffer.
17. Elute the chromatin from beads by incubation with 500 μl of ChIP elution buffer at 65 °C for 30 min. For ChIP-seq, eluting chromatin at 95 °C should be avoided as DNA will be denatured. Subsequent annealing of DNA duplexes from a complex genome may lead to irregular DNA ends and spurious annealing of repetitive sequences may result.
18. Pellet the beads and transfer the supernatant into a new tube. Add 20 μl of 5 M NaCl into each tube and incubate at 65 °C overnight to reverse the cross-linking. We strongly suggest performing the reverse cross-linking for more than 12 h because samples with incomplete reversal of cross-linked chromatin could be refractory to Covaris S2 fragmentation.
19. On the next day, add 20 μg of proteinase K into each sample and incubate at 45–65 °C for 1–2 h.
20. Extract the ChIPed DNA with Phenol/Chloroform/isoamyl alcohol (25:24:1) and transfer the aqueous phase into a clean tube. Add 1 μl of glycogen, 50 μl of 3 M NaOAc (pH 5.3), and 1 ml of 100 % ethanol into the tube and centrifuge at 17,200 × *g* for 10 min. Wash with 80 % ethanol and air dry the pellet before resuspend with 30 μl of $H_2O$.
21. Purify the ChIPed DNA once again with MinElute Reaction Cleanup Kit to completely remove the residual SDS and salts. Elute the DNA in 30 μl and take 2 μl to make a 1:10 dilution for evaluating the quality of the ChIP process.

### 3.2 Validate the Quality of ChIP Before Constructing Sequencing Libraries

It is necessary to verify the quality of ChIP before proceeding into library construction and sequencing as these steps are frequently time consuming and costly. The quality of ChIP can be evaluated by quantitative PCR targeting regions that chromatin states are known. For example, transcription start sites of actively expressed genes such as *TUB8* or *GAPDH* are known to associate with H3K4me3 or H3K9Ac, whereas the tightly regulated *FLC* or *STM* loci are enriched in the H3K27me3 mark. It is important that minimal DNA template is detected for the mock ChIP control to ensure the specificity of ChIP-seq results. Washing the Protein A agarose beads for more rounds or increasing the concentration of NaCl up to 500 mM in the ChIP dilution buffer may help to reduce nonspecific binding between chromatin and beads.

### 3.3 Construction of SOLiD Sequencing Libraries

#### 3.3.1 Preparation of Adaptors for Library Construction

1. Dissolve the synthesized oligos with nuclease-free water to a final concentration of 100 μM.
2. Mix oligos corresponding to each strand of the adaptor with equal amount in a PCR tube. Add 5× or 10× PCR buffer into the oligo mixture to adjust the solution condition equivalent to 1× PCR buffer. The ionic strength of the 1× PCR buffer should be sufficient to facilitate the annealing of DNA.
3. Incubate the oligo mixture in a thermo-cycler with the following program to hybridize the oligonucleotide pairs:

| | |
|---|---|
| 95 °C | 5 min |
| 72 °C | 5 min |
| 60 °C | 5 min |
| 50 °C | 3 min |
| 40 °C | 3 min |
| 30 °C | 3 min |
| 20 °C | 3 min |
| 10 °C | 3 min |
| 4 °C | ∞ |

   As the oligos for P1 adaptors were synthesized at PAGE purified grade, it is unnecessary to gel purify the annealing P1 adaptor. However, gel purification is essential for P2 adaptors as the oligos are synthesized at the desalted grade and therefore contain large amount of incompletely synthesized oligos.

4. Run 10 μg of the each P2 barcoded adaptors on a 3 % agarose gel and stain the gel with ethidium bromide or reduced-toxicity alternative after finishing the electrophoresis.
5. Gel purify the predominant 50 nucleotide band with the QIAquick Gel Extraction kit and elute in 30 μl.
6. Adjust the concentration of P1 and P2 adaptor to 250 and 500 ng/μl, respectively.
7. Store the gel purified adaptor at −20 °C until ready for the ligation (*see* **Note 2**).

### 3.3.2 Fragment the ChIPed DNA with Covaris S2

The fragmentation of chromatin with a probe sonicator, such as Branson S150, will generate chromatin fragment with a size range of 300–500 bp. The optimal insertion size of SOLiD™ 2.0 library is around 200 bp. Therefore the ChIPed DNA needs to be further fragmented by the Covaris S2 before ligation with appropriate adaptors.

1. Dilute each ChIPed DNA sample to 300 μl with water and transfer the entire sample into a T6 round bottom glass tube. Fill the tube with water to prevent the formation of air bubble during the Covaris treatment and snap on the cap.
2. Mount the tube onto a Covaris S2 instrument and run the following program:

| Treatment 1 for 5 s | |
|---|---|
| Duty cycle | 0.5 % |
| Intensity | 8 |
| Cycles/burst | 50 |
| *7 cycles* | |
| Treatment 2 for 60 s | |
| Duty cycle | 20 % |
| Intensity | 8 |
| Cycle/burst | 200 |

3. Transfer the sample into a clean microcentrifuge tube with an extended gel loading tip.
4. Precipitate the DNA with 1 μl of glycogen, 1/10 volume of NaOAc and 2 volume of 100 % ethanol. Wash with 80 % ethanol and air dry the pellet before resuspend with 34 μl of $H_2O$.

### 3.3.3 Ligation with P1 and Barcoded P2 Library Adaptors

1. Carry out enzymatic reaction to convert the ends of sonicated DNA to phosphorylated blunt-ends following the manual of End-It™ DNA End-Repair Kit using all of the 34 μl of DNA after the Covaris treatment. Incubate the reaction at room temperature for 45 min.
2. Purify the DNA with MinElute Reaction Cleanup Kit. Elute with 20 μl of EB buffer as supplied in the kit.
3. Set up the following ligation reaction for each library to be made.

| 2 μl | 10× Ligation Buffer |
|---|---|
| 1 μl | ATP (10 mM) |
| 1 μl | P1 adaptor (250 ng/μl) |
| 1 μl | Barcoded P2 adaptor (500 ng/μl) |
| 1 μl | T4 DNA Ligase |
| 10 μl | sample DNA |
| 4 μl | $H_2O$ |

Incubate the ligation reaction at room temperature for 1 h.

4. Extract the ligation reaction with Phenol/Chloroform/IAA (25:24:1) to remove the DNA ligase. Ligase will interfere with the migration of DNA during electrophoresis.
5. Precipitate the ligation product with 1 μl of glycogen, 1/10 volume of NaOAc, and 2 volume of 100 % ethanol. After washing once with 80 % ethanol, resuspend the pellet in 10 μl of $H_2O$ that can entirely be loaded into a well of the PAGE gel.
6. Prepare a 6 % native polyacrylamide TBE gel for the size selection of ligation products with apparatus for regular SDS-PAGE gels as follows:

| | |
|---|---|
| 1 ml | 40 % Acrylamide/Bis (19:1) solution |
| 700 μl | 5× TBE |
| 20 μl | TEMED |

Add $H_2O$ to adjust the total volume to 7 ml.

Add 50 μl of 10 % fresh ammonium persulfate (APS) and mix the gel solution. Pour the gel immediately and let it sit for 10 min to solidify.

7. Load the whole ligation product into one lane of the polyacrylamide gel. Use a 50 bp DNA ladder as the size reference. Run the gel at 120 V until the bromophenol blue marker reached near the bottom of the gel.
8. Stain the gel with ethidium bromide or reduced-toxicity alternative and excise the gel piece containing 150–200 bp fragments on a UV-light box.
9. Vertically excise the gel piece into three slices to use one of them for the determination of cycle numbers for library amplification. It is important that the gel piece is sliced vertically rather than horizontally to ensure that DNA contained in the three pieces have identical size distribution. Gel slices can be stored at −20 °C until ready to use.

#### 3.3.4 Library Amplification

There are two reasons to perform library amplifications during the preparation of sequencing libraries. (1) The size-selected ligation products contain both ligated and unligated DNAs; library amplifications can enrich DNA molecules that are ligated with P1 and P2 adaptor at each end because only these molecules can serve as template during the PCR reactions. (2) Library amplification is essential for generating enough library DNA to allow accurate quantification with spectrophotometry. The efficiency of emulsion PCR to prepare beads for SOLiD™ sequencing is highly sensitive to the quantity of library DNA added into the reaction. However, overamplification of the library would be detrimental to their complexity.

Therefore the cycle numbers need to be empirically tested for each library to minimize over-amplification while providing enough DNA for quantification. For the ChIP-seq libraries, we have prepared using 0.5–1.5 g of *Arabidopsis* leaf tissues, we found the optimal number of PCR cycles to be between 14 and 20.

1. Transfer one gel slice prepared in **step 9** of Subheading 3.3.3 into a PCR tube. Crush the gel slice with a 200 μl pipetting tip.
2. Set up a PCR reaction with the crushed gel slice as template.

| | |
|---|---|
| 25 μl | FailSafe PCR 2× PreMix E |
| 1 μl | P1 primer (100 μM) |
| 1 μl | P2 primer (100 μM) |
| 1 μl | FailSafe PCR Enzyme Mix |
| 22 μl | $H_2O$ |

Perform a PCR reaction with the following condition on a thermal cycler.

| | |
|---|---|
| 72 °C | 20 min (*see* **Note 3**) |
| 95 °C | 2 min |
| 95 °C | 15 s |
| 62 °C | 15 s |
| 72 °C | 1 min (14–20 cycles) |
| 72 °C | 5 min |
| 4 °C | ∞ |

3. Starting from the extension step of cycle 14, pause the thermal cycler 2–3 s before the extension step finish and transfer 10 μl of the reaction into a clean tube. Do not transfer the PCR reaction at the denaturing step, as DNA may not anneal properly when suddenly exposed to room temperature. Repeat this procedure every two cycles until the PCR reaction is completed.
4. Run all the PCR products after different cycles of amplification on a 6 % native polyacrylamide gel and visualize through ethidium bromide staining or reduced-toxicity alternative (Fig. 1). Choose the lowest cycle number which shows a robust library product appearing at around 200 bp for the library amplification (e.g., cycle 16 in the example shown in Fig. 1).
5. Perform full library amplifications with the cycle number selected for each library.
6. Ethanol-precipitate the whole PCR reaction and resuspend the pellet in 20 μl water. Load the entire product onto one lane of

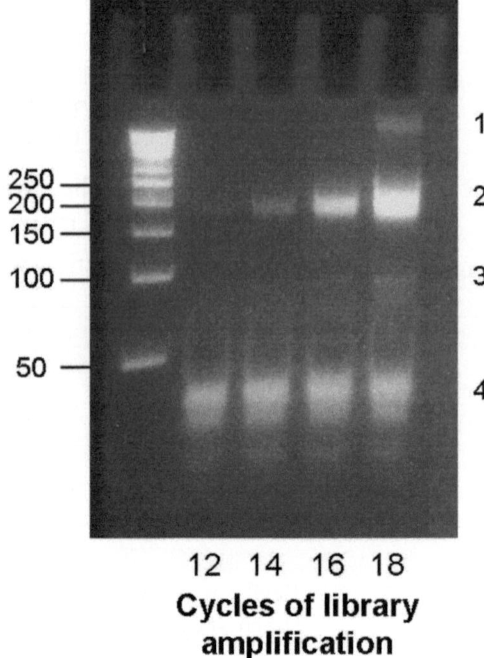


**Fig. 1** The empirical determination of cycle numbers for the library amplification. Products of 12–18 cycles of library amplification were resolved on a 6 % native polyacrylamide gel. Labels 1–4 indicate possible products that can be observed on the gel—(*1*) nonspecific band presumably generated through annealing between products. This band appears when the library was amplified for excess cycles and should be avoided; (*2*) the library product; (*3*) presumable adaptor dimers (*see* **Note 2**); (*4*) primers that have not been incorporated

a 3 % agarose gel and purify the band around 200 bp with QIAquick gel extraction kit. Elute the libraries in 15 μl to maximize the DNA concentration for accurate quantifications.

7. Use 1 μl of each of the purified libraries to determine their concentration with a Nanodrop spectrophotometer.

### 3.4 Evaluate the Effect of Read Length on the Alignment to the Arabidopsis Genome

The read length of sequencing tags is critical for the accuracy of aligning tags onto the reference genome. Shorter tags are more likely to match multiple positions in a genome and thus decrease the fidelity of the alignment. Longer tags, instead, would linearly increase the cost of sequencing while they may not necessarily improve the confidence of the alignment. In addition, unlike RNA-seq where longer reads could provide useful information such as splice site variations, tag sequences in a ChIP-seq experiment would be identical with the reference genome and longer reads than necessary for unambiguous alignment are usually of little value. The minimal read length that can generate satisfying alignment results is dependent on multiple factors including the genome size and the amount of repetitive sequences in the genome.

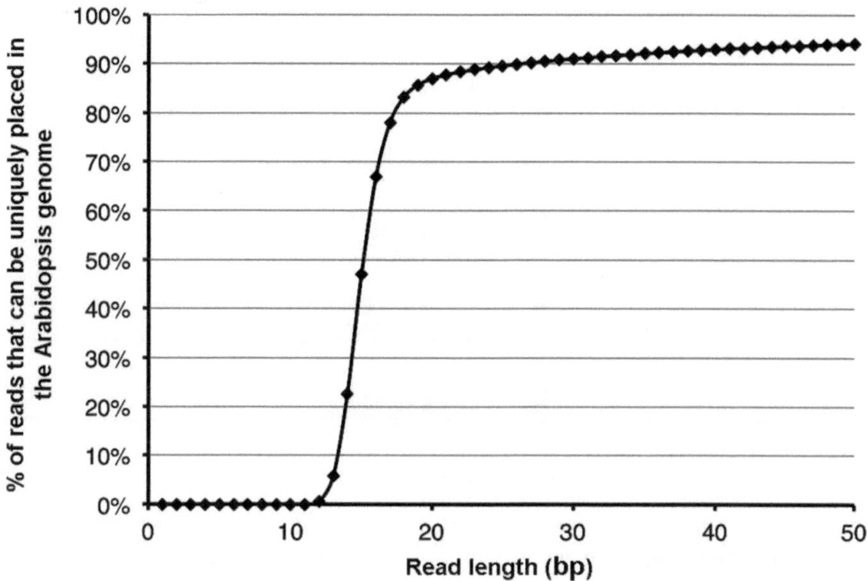


**Fig. 2** Evaluation of the effect of read length on alignment with the *Arabidopsis* genome. The percentage of reads that can be matched to a unique position in the genome was calculated for reads with length from 1 to 50 bp

Therefore it is essential to determine the relation between read length and alignment accuracy for each genome of interest in order to optimize the target tag sequence length in a ChIP-seq study.

To estimate the read length that is sufficient for aligning tags to the *Arabidopsis* genome, we plotted the read length versus the percentage of reads that is unique in the genome (Fig. 2). We found that more than 90 % of all possible 26 nucleotide (nt) reads can be uniquely placed in this genome and the increase of this percentage is relatively lower for reads longer than 26 nt. Notably, for reads that are 50 nt long, there are still 5.9 % of reads that are not unique in the genome. This analysis shows that sequence reads with length between 30 nt (91 % of which are unique) and 35 nt (92.1 % of which are unique) would be suitable for ChIP-seq with the *Arabidopsis* genome (*see* **Note 4**).

### *3.5 Evaluating the Impact of Sequencing Depth on the Profile of Chromatin Modifications*

The sequencing depth that will be needed to quantitatively profile a chromatin modification is dependent on two factors. (1) The size of the reference genome. A larger reference genome such as human or maize genome would require more tags to cover compared to smaller genomes such as that of *S. cerevisiae* or *Arabidopsis.* (2) The pattern of the chromatin modification being profiled. For example, profiling a spatially focused chromatin mark such as H3K4me3 that clusters around transcription start sites would not require many tags even for a large genome. In contrast, ChIP-seq of histone H3 for measuring genome-wide nucleosome density might need a much large number of tags to avoid sampling biases.

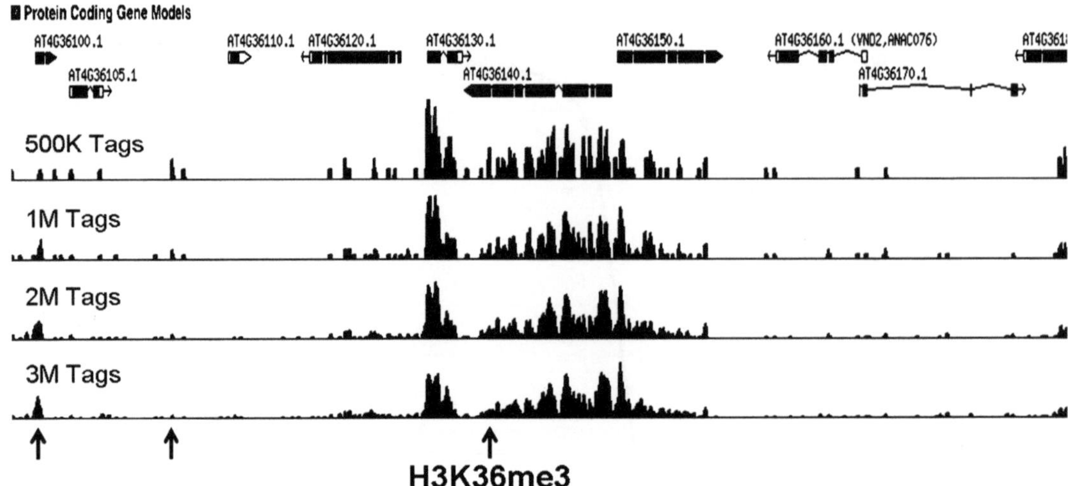


**Fig. 3** Evaluation of the effect of sequencing depth on the profiled pattern of histone modification for the H3K36me3 mark. Coverage of this genomic region by sequencing tags (H3K36me3 ChIP) is represented by "coverage per million tags." The profiles of H3K36me3 were generated by using 500 K, 1 million, 2 million, and 3 million of sequenced tags. The scales of the *y*-axis are identical across the panels

We have empirically evaluated the impact of sequencing depth on the resulting pattern of chromatin modification. For a region of *Arabidopsis* chromosome 4, we compared the pattern of H3K36me3 that is profiled with 500 K, 1 million, 2 million, and 3 million tags (Fig. 3). The profiles have been scaled and presented as "coverage per million tags" in order to compensate the differences caused solely by total tag numbers. Except for a few minor differences as indicated by arrows in Fig. 3 and the "noisier" outline of pattern in the 500 K tag dataset, the patterns are essentially identical between the four panels. Similar results were obtained by analyzing other genomic regions or histone modifications (data not shown). Therefore, we suggest that 2–3 million of good quality tags are usually sufficient to quantitatively profile a specific histone modification in the *Arabidopsis* genome.

### 3.6 Comparison Between ChIP-seq Profiles and Published ChIP-chip Profiles

Several histone modifications have been profiled globally with ChIP-chip using the Affymetrix *Arabidopsis* Tiling 1.0 array. The profiles of H3K4me3 and H3K27me3 generated from published ChIP-seq results were compared with those generated by ChIP-chip (Fig. 4, [12]). Both work used aerial tissues from 2-week-old *Arabidopsis* plants. We plotted the pattern of H3K4me3 and H3K27me3 marks as profiled by ChIP-seq or ChIP-chip in a 50 kb region surrounding the *AGAMOUS* locus. The profiles generated by the two techniques are highly similar and "sharper" ChIP-seq signals can be found at every H3K4me3 or H3K27me3 peak that are detected with ChIP-chip, which suggests that ChIP-seq effectively captured the patterns that can be identified by ChIP-chip.

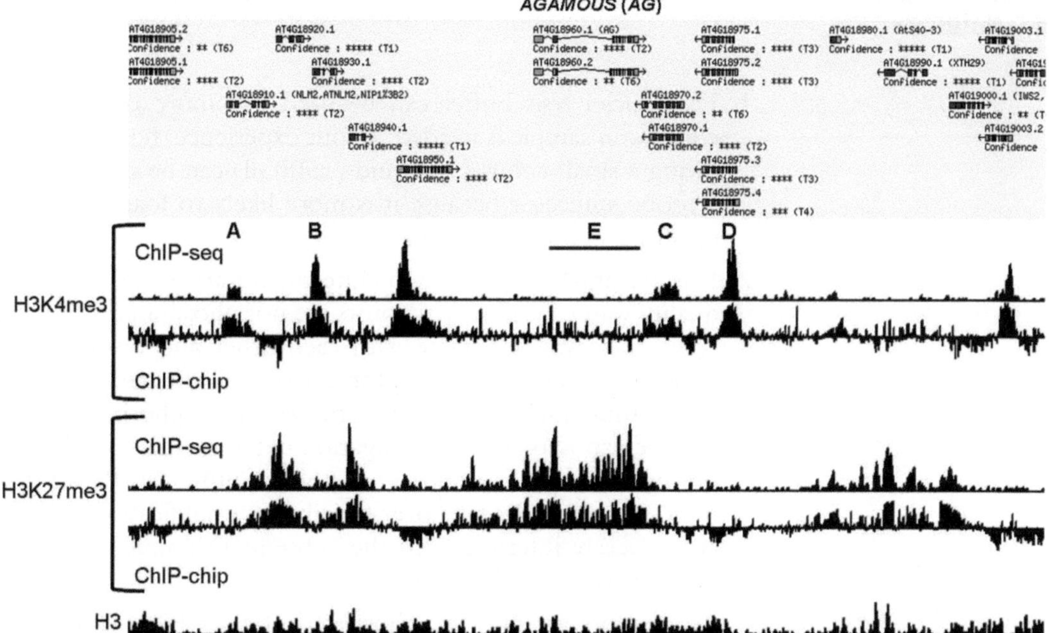


**Fig. 4** Comparison of H3K4me3 and H3K27me3 profiles generated with ChIP-seq and ChIP-chip. H3K4me3 and H3K27me3 patterns profiled with ChIP-seq and ChIP-chip were plotted for a 50 kbp region surrounding the *Arabidopsis AGAMOUS* locus. Signals in ChIP-seq and ChIP-chip panels each indicate coverage by sequencing tags and log2 values (H3K4me3 or H3K27me3/H3), respectively. An H3 ChIP pattern produced by ChIP-seq was shown to control the nucleosome density in this region

ChIP-seq may allow more accurate quantifications as shown by the differential enrichments of H3K4me3 between peaks A and B (Fig. 4). A ~3-fold difference regarding the amplitude of peaks was detected between peaks A and B with ChIP-seq, whereas the two peaks were shown to be nearly equal when assayed with ChIP-chip even considering that the ChIP-chip results were presented in a logarithm form. A similar case is also shown by comparing peaks C and D. In addition, ChIP-seq results apparently showed greater signal to noise ratios as the quantification is not affected by background florescence signals as is the case for microarrays. For example, region E corresponds to the transcribed region of *AGAMOUS* and is enriched with the H3K27me3 mark, which should mean that it is devoid of the H3K4me3 mark. Although few tags were found in region E for the H3K4me3 mark by ChIP-seq, ChIP-chip data for H3K4me3 has yielded a substantial amount of sporadic signal in this region. Overall, although we cannot rule out that different antibodies and plant growth conditions cause the observed distinctions between ChIP-seq and ChIP-chip results, we can nevertheless conclude that ChIP-seq is able to generate data with quality equal to or better than the ChIP-chip technique.

## 4 Notes

1. Less nuclei lysis buffer can be used if a more concentrated chromatin sample is needed. In our experience, however, sonicating a small volume of liquid (<500 μl) can be difficult with a probe sonicator because it is more likely to foam as well as resulting in more heating.
2. It is extremely important to elute the library adaptors in high quality water. As library adaptors are not phosphorylated, generally they will not ligate with each other and form adaptor dimers. However, if the adaptors are partially degraded by contaminating nuclease activities, the terminal phosphor group may be exposed and enable ligation between adaptors. Due to the excess amount of adaptors, the resulting P1–P2 adaptor dimers could easily take over the library amplification reaction and severely interfere with the subsequent amplification and gel purification of libraries.
3. As adaptors are not phosphorylated at their 5′-end, only one strand of the adaptor can form the diester bond with the insertion DNA whereas a nick will remain in the other strand. This nick needs to be repaired by the 5′→3′ exonuclease activity of Taq polymerase before the DNA is denatured in the pre-PCR step.
4. The scripts for performing the analysis can be downloaded at http://aesop.rutgers.edu/~lamlab/resources/resources.html

## Acknowledgments

We thank Drs. R.A. Kerstetters, T.P. Michael, D. Sidote, and M. Diamond at the Waksman Genomic Core Facility for their technical support and suggestions in developing the ChIP-seq procedure.

### References

1. Mito Y, Henikoff JG, Henikoff S (2005) Genome-scale profiling of histone H3.3 replacement patterns. Nat Genet 37: 1090–1097
2. Goldberg AD, Banaszynski LA, Noh KM, Lewis PW, Elsaesser SJ, Stadler S, Dewell S, Law M, Guo X, Li X, Wen D, Chapgier A, DeKelver RC, Miller JC, Lee YL, Boydston EA, Holmes MC, Gregory PD, Greally JM, Rafii S, Yang C, Scambler PJ, Garrick D, Gibbons RJ, Higgs DR, Cristea IM, Urnov FD, Zheng D, Allis CD (2010) Distinct factors control histone variant H3.3 localization at specific genomic regions. Cell 140:878–891
3. Barski A, Cuddapah S, Cui K, Roh TY, Schones DE, Wang Z, Wei G, Chepelev I, Zhao K (2007) High-resolution profiling of histone methylations in the human genome. Cell 129:823–827
4. Wang Z, Zang C, Rosenfeld JA, Schones DE, Barski A, Cuddapah S, Cui K, Roh TY, Peng W, Zhang MQ, Zhao K (2008) Combinatorial patterns of histone acetylations and methylations in the human genome. Nat Genet 40:897–903
5. Bernstein BE, Mikkelsen TS, Xie X, Kamal M, Huebert DJ, Cuff J, Fry B, Meissner A, Wernig M, Plath K, Jaenisch R, Wagschal A, Feil R,

Schreiber SL, Lander ES (2006) A bivalent chromatin structure marks key developmental genes in embryonic stem cells. Cell 125: 315–326

6. Mikkelsen TS, Ku M, Jaffe DB, Issac B, Lieberman E, Giannoukos G, Alvarez P, Brockman W, Kim TK, Koche RP, Lee W, Mendenhall E, O'Donovan A, Presser A, Russ C, Xie X, Meissner A, Wernig M, Jaenisch R, Nusbaum C, Lander ES, Bernstein BE (2007) Genome-wide maps of chromatin state in pluripotent and lineage-committed cells. Nature 448:553–560
7. Deal RB, Henikoff S (2010) A simple method for gene expression and chromatin profiling of individual cell types within a tissue. Dev Cell 18:1030–1040
8. Zhang X, Yazaki J, Sundaresan A, Cokus S, Chan SW, Chen H, Henderson IR, Shinn P, Pellegrini M, Jacobsen SE, Ecker JR (2006) Genome-wide high-resolution mapping and functional analysis of DNA methylation in *Arabidopsis*. Cell 126:1189–1201
9. Applied Biosystems SOLiD™ System 2.0 User Guide (2008)
10. Park PJ (2009) ChIP-seq: advantages and challenges of a maturing technology. Nat Rev Genet 10:669–680
11. Applied Biosystems SOLiD™ Multiplexing Protocol (2008)
12. Oh S, Park S, van Nocker S (2008) Genic and global functions for Paf1C in chromatin modification and gene expression in *Arabidopsis*. PLoS Genet 4:e1000077

# Chapter 13

# Analysis of Retrotransposon Activity in Plants

## Christopher DeFraia and R. Keith Slotkin

## Abstract

Retrotransposons are transposable elements that duplicate themselves by converting their transcribed RNA genome into cDNA, which is then integrated back into the genome. Retrotransposons can be divided into two major classes based on their mechanism of transposition and the presence or absence of long terminal repeats (LTRs). In contrast to mammalian genomes, in which non-LTR retrotransposons have proliferated, plant genomes show evolutionary evidence of an explosion in LTR retrotransposon copy number. These retrotransposons can comprise a large fraction of the genome (75 % in maize). Although often viewed as molecular parasites, retrotransposons have been shown to influence neighboring gene expression and play a structural and potential regulatory role in the centromere. To prevent retrotransposon activity, eukaryotic cells have evolved overlapping mechanisms to repress transposition. Plants are an excellent system for studying the mechanisms of LTR retrotransposon inhibition such as DNA methylation and small RNA-mediated degradation of retrotransposon transcripts. However, analysis of these multi-copy, mobile elements is considerably more difficult than analysis of single-copy genes located in stable regions of the genome. In this chapter we outline methods for analyzing the progress of LTR retrotransposons through their replication cycle in plants. We describe a mixture of traditional molecular biology experiments, such as Southern, Northern, and Western blotting, in addition to nontraditional techniques designed to take advantage of the specific mechanism of LTR retrotransposition.

**Key words** Transposable element, Retrotransposon, LTR, Epigenetic regulation, DNA methylation, Transposition

## 1 Introduction

Transposable elements make up a large fraction of plant genomes, comprising 14 % of the small genome of *Arabidopsis*, and 85 % of the much larger genome of maize [1, 2]. The majority of these elements in plant genomes are long terminal repeat (LTR) retrotransposons, which comprise 9 % of the genome in *Arabidopsis* [1] and 75 % in maize [2]. Despite their abundance, most LTR retrotransposons are not full-length autonomous elements, as all retrotransposons continuously generate defective derivative copies [3, 4]. In addition, the plant cell uses multiple overlapping mechanisms of

Charles Spillane and Peter C. McKeown (eds.), *Plant Epigenetics and Epigenomics: Methods and Protocols*, Methods in Molecular Biology, vol. 1112, DOI 10.1007/978-1-62703-773-0_13, 

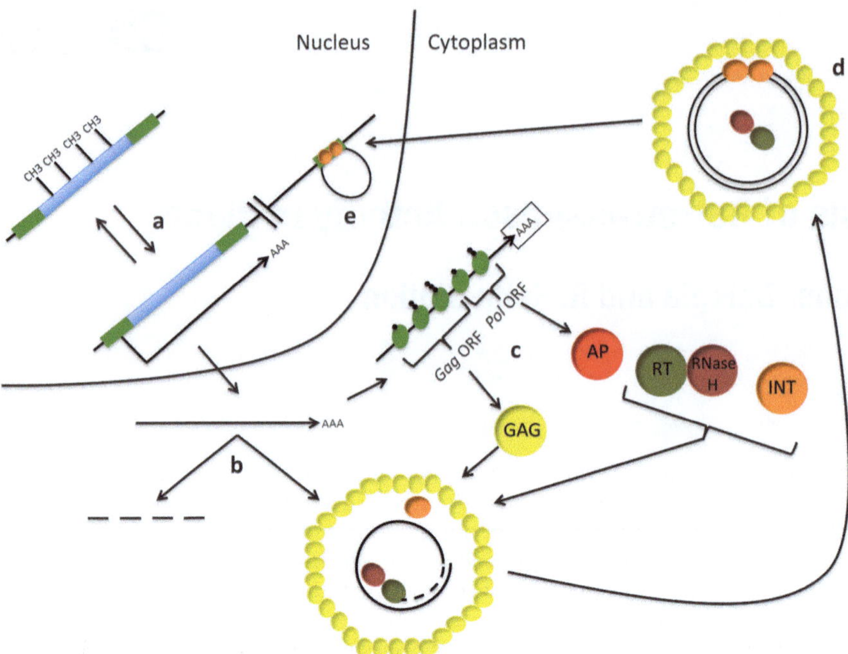


**Fig. 1** Replication cycle of an LTR retrotransposon. (**a**) The chromatin-level regulation of an LTR retrotransposon can be determined by analysis of DNA methylation states and histone tail modifications. (**b**) Transcription and fate of the retrotransposon transcripts are tested by (q)RT-PCR, Northern blot, and GUS enhancer traps. (**c**) Retrotransposon protein accumulation and activity are tested by Western blot, reporter fusions, and PERT assay. (**d**) The formation of nonchromosomal retrotransposon cDNA intermediates are tested for by Southern blot and ligation-mediated PCR. (**e**) Finally, integration of LTR retrotransposon copies into the host genome is tested by Southern blot and transposable element display. Figure adapted from Sabot and Schulman [42]

repression against transposable element duplication, resulting in the inhibition of LTR retrotransposon activity [5].

Repression of LTR retrotransposon activity can take many forms. Therefore, when chemical treatments, stress treatments, or mutations are evaluated for their effect on retrotransposon activity, it is important to determine which step of the LTR retrotransposon replication cycle is affected. For example, DNA methyltransferase mutants in *Arabidopsis* display transcriptional activation of retrotransposons [6, 7], but these mutations do not result in LTR retrotransposon mobilization, demonstrating that there are subsequent levels of LTR retrotransposon regulation.

Figure 1 represents the replication pathway for a stereotypical plant LTR retrotransposon and depicts the points of retrotransposon repression. Most plant LTR retrotransposons are primarily inhibited at the transcriptional level by epigenetic silencing, manifested as chromatin modifications such as DNA methylation, condensation of nucleosomes, and the post-translational modification of histone tails [8]. When expressed, the LTR retrotransposon transcripts can potentially be degraded in the

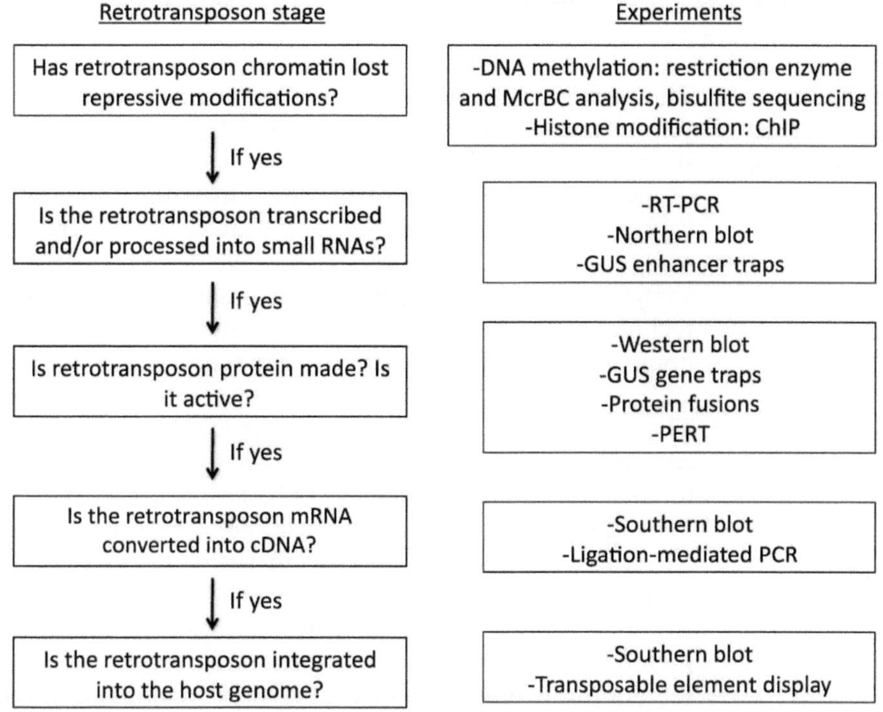


**Fig. 2** Hierarchical experimental guide for determining the effect of a mutation or treatment on the life cycle of a retrotransposon. The suggested order of inquiry into the retrotransposon life cycle is based on the fact that each step in the life cycle requires completion of the previous step

cytoplasm post-transcriptionally by small RNA mediated transcript cleavage [9]. The reverse transcription step and the translation of LTR retrotransposon proteins may also be inhibited, as well as nuclear import and element integration in the genome.

To obtain a comprehensive understanding of the effects of removing LTR retrotransposon inhibition, and thus understand LTR retrotransposon activity and host repression, each step of the LTR retrotransposon duplication cycle can be analyzed. In this chapter the experimental approaches used to test each stage of the retrotransposon life cycle are outlined, including loss of chromatin-level regulation, retrotransposon expression, retrotransposon protein accumulation and activity, accumulation of retrotransposon cDNA copies, and finally transposition. A hierarchical workflow is presented in Fig. 2 to help guide experiments aimed at dissecting the various levels of LTR retrotransposon activity as well as the repression mechanisms by the plant. This workflow can be applied to identify the specific effect of a given treatment or mutation on the LTR retrotransposon replication pathway. While this chapter provides a guide for investigating the regulation and transposition cycle of LTR retrotransposons, it does not provide experimental details. The reader is therefore encouraged to consult the publications cited herein for further details, protocols, and additional information.

## 2 Methods

### 2.1 Does the LTR Retrotransposon Lose Repressive Chromatin Modifications?

The initial layer of repression of LTR retrotransposons in plants is on the transcriptional level, mediated by chromatin modification such as DNA methylation and/or post-translational modification of histone tails [5]. This section describes three methods for analyzing retrotransposon DNA methylation and one technique for analyzing histone tail modifications. We describe methods for the analyses of single and multiple retrotransposons. However, these techniques can be scaled to the genome-wide level by generating libraries and subjecting them to deep sequencing [10, 11].

#### 2.1.1 Analysis of LTR Retrotransposon DNA Methylation

##### Analysis of DNA Methylation by Restriction Enzymes

The traditional method for analyzing DNA methylation is to use restriction enzymes that are sensitive to cytosine DNA methylation in their recognition sequence [12, 13]; see also the chapter of Parisod et al. elsewhere in this volume. Genomic DNA restricted with these enzymes will only be efficiently digested if the recognition sequence is not methylated. A list of restriction enzyme sensitivities to cytosine DNA methylation can be found at http://rebase.neb.com/rebase/rebms.html. Cleaved DNA can be used for Southern blotting and probed with regions of retrotransposons, as in Cao and Jacobsen [14]. This analysis can be performed with or without knowledge of the sequence context of the genomic regions being assayed as long as the sequence or identity of the probe is known. If the sequence surrounding the retrotransposon is known, the DNA digested with the methylation-sensitive restriction enzyme can be used for PCR with primers surrounding the restriction site(s) to assay cleavage efficiency. For techniques related to analyzing differential methylation by AFLP, see the chapter by Parisod et al. elsewhere in this volume.

In plants, cytosines are methylated in CG, CHG (where H is any base except G), and CHH sequence contexts. Distinct silencing pathways are associated with each of these methylation contexts [15]. LTR retrotransposons are targets of all of these methylation types [11, 16]. For analyzing DNA methylation patterns using traditional restriction enzymes, particular enzyme pairs designed to investigate methylation in one specific sequence context can be employed [17]. For example, digestion with the restriction enzyme *HpaII* tests for DNA methylation in a CG sequence context, while the methylation-insensitive enzyme *MspI* can be used as a control to assay efficient digestion of the DNA. Digestion of genomic DNA using methylation-sensitive restriction enzymes provides semiquantitative data on how much of the restriction target site is methylated. However this technique only provides information on one to several methylation sites at a time.

### Analysis of DNA Methylation by McrBC-PCR

McrBC-PCR is the most rapid and inexpensive method to test whether multiple retrotransposons are methylated. In this assay, genomic DNA is subjected to digestion with the McrBC enzymes. These enzymes work in tandem to cleave only DNA that contains methylated cytosine bases, with very little sequence specificity. Following digestion, the DNA is used directly for PCR with retrotransposon-specific PCR primers as in Lippman et al. [18]. As methylated regions of DNA will be digested into small fragments, these regions will fail to amplify, whereas unmethylated DNA will not be cleaved and, therefore, amplify. End-point PCR can be performed on the cleaved DNA, which provides a semiquantitative measure of DNA methylation. If a more quantitative measure is required, real-time quantitative PCR, or (q)PCR, may be employed.

In both approaches described above, several controls are necessary to ensure accurate interpretation of results. First, a mock-digested DNA sample must be included alongside the digested DNA to ensure that the DNA can be efficiently amplified and to provide a baseline for the detection of undigested DNA. In addition to amplifying or probing the region of interest, both a region of DNA known to be heavily methylated and a region of DNA known to lack significant methylation should be assayed. The methylated control, which should not efficiently amplify after digestion, serves as a positive control to confirm the efficiency of the digestion, while the unmethylated control, which should efficiently amplify, serves as a negative control and confirms that the input DNA is intact and suitable for PCR.

McrBC-PCR is rapid, inexpensive, and does not require much technical expertise. However the information it yields about DNA methylation status is limited. Since McrBC cleaves near multiple methylated cytosines in any given sequence, it cannot be used to distinguish between methylation in different sequence contexts (CG, CHG, or CHH). Additionally, because the readout of this assay is the quantity of amplified DNA, it does not provide single cytosine nucleotide resolution. Therefore McrBC-PCR represents a rapid way to rapidly examine the methylation of multiple retrotransposons. However, the assay provides poor resolution of the methylation sequence context.

### Analysis of DNA Methylation by Bisulfite Sequencing

Bisulfite sequencing involves treatment of single-stranded DNA with bisulfite, which converts unmethylated cytosines, but not methylated cytosines, into uracil. During PCR amplification of bisulfite-treated DNA, the uracil bases are replaced by thymine bases. The amplified DNA is then sequenced to determine which cytosines were converted (unmethylated) and which cytosines were not converted (methylated). Several protocol kits are commercially available for bisulfite conversion of DNA and subsequent purification. The design of primers for this amplification is a key step and requires sequence knowledge of the target DNA. Since, in plants, any cytosine on the template DNA may be converted to uracil, a mixture of primers containing either a C or a T at a given position is used. Publically available software for the design of bisulfite primers can assist in this process. One such web-based tool, Kismeth, was designed specifically for the design of primers and analysis of data from bisulfite sequencing experiments in plants [19]. Since there are typically many copies per family of LTR retrotransposons in plant genomes, primers may be designed to amplify all or a subset of these elements based on the complementarity and specificity of the primers.

Obtaining a single PCR amplicon from bisulfite-treated DNA is the most difficult step in bisulfite sequencing. Bisulfite treatment can cause DNA damage and inhibit PCR. Touchdown PCR and the use of primer mixtures described above can facilitate amplification. Optimization of PCR conditions may be necessary to obtain enough amplified DNA for cloning and sequencing. To facilitate sequence analysis, sequencing data can be imported into CyMATE (http://www.cymate.org/) or Kismeth (http://katahdin.mssm.edu/kismeth/revpage.pl) for visualization of methylation patterns in all sequence contexts. If amplifying fragments from high copy retrotransposons, sequences can also be analyzed by ClustalX (http://www.clustal.org/) to determine which elements were sequenced, provided sufficient polymorphism exists to distinguish between individual copies.

In any bisulfite sequencing experiment, a region of DNA known to be unmethylated must be examined to ensure efficient bisulfite conversion of non-methylated cytosines (typically above 97 %). Since methylation of a given cytosine may vary from cell to cell, several clones from each sample are sequenced [20]. Results may be expressed as a dot blot, which portrays each cytosine as an open (unmethylated) or filled (methylated) circle. Alternatively, the percentage of methylated cytosines at each position can be displayed [21]. For both figures, different colored circles or bars can represent the different sequence contexts of each cytosine.

Bisulfite sequencing is more laborious than using methylation-sensitive restriction enzymes or McrBC and requires more technical expertise. It can, however, yield single base resolution of methylation patterns and be quantitative, provided sufficient sequences are examined from each sample.

#### 2.1.2 Analysis of LTR Retrotransposon Histone Tail Modifications

Modifications on the N-terminal tails of histone proteins coordinate the chromatin-level control of the associated DNA. In plant genomes, there is a tight interconnection between DNA methylation and histone tail modification [7, 22]. For example, when DNA methylation is lost, histone tail modifications are altered. Conversely, alterations in histone tail modifications result in changes in DNA methylation.

Transcriptionally repressed chromatin, which is often associated with LTR retrotransposon DNA, is typically enriched in histone H3 methylated at lysine 9 (meH3K9) [18, 23], and/or lysine 27 [24]. In contrast, genes and transcriptionally active LTR retrotransposons have reduced methylation at lysine 9 and 27 and are associated with H3 methylation at lysine 4 (meH3K4). Each of these modifications can be assayed using meH3K4-, meH3K9-, or meH3K27-specific antibodies with chromatin immunoprecipitation (ChIP); see the chapters of Song et al. and of Weinhofer & Köhler, elsewhere in this volume for further applications of ChIP to plant biology. The commercial availability of a wide range of ChIP antibodies has been facilitated by the fact that the N-terminal portion of histone tails is highly conserved. Therefore, antibodies raised against animal histone proteins cross-react with plant histones.

A ChIP experiment requires multiple experimental and internal controls [25]. First, DNA and its associated proteins are cross-linked together, and this chromatin complex is isolated. Next the chromatin is fragmented into roughly the size of one to a few nucleosomes. Some of this input chromatin is set aside and used to control for differences in the amount of PCR-competent DNA in each sample. The chromatin containing the modified histone of interest is then immunoprecipitated with the appropriate primary and secondary antibodies. A control lacking the histone-specific primary antibody should be included to detect any spurious pull-down of chromatin of interest and determine the background level of immunoprecipitation. Pre-blocking of the primary and second ary antibodies with non-plant DNA or chromatin and the use of siliconized tubes reduces this background. The immunoprecipitated chromatin is un-cross-linked and subjected to semiquantitative PCR or (q)PCR to specifically amplify short regions of the genome. Similar to the McrBC analysis described above, design of these PCR primers requires some prior knowledge of the LTR retrotransposon sequence. Fragments of the genome with known histone modification patterns should be tested to confirm the specificity of the assay.

ChIP is a powerful technique that can be used to determine the association of an individual LTR retrotransposon copy with a histone mark by using element-specific PCR primers. Alternatively, entire families can be simultaneously assayed using multi-target primer sets. In addition, ChIPed DNA can be scaled for deep sequencing, producing genomic maps of histone modification patterns [11].

### 2.2 Fate of LTR Retrotransposon Transcripts

Once the chromatin-level inhibition of LTR retrotransposons is released, the retrotransposon will likely become transcriptionally active, generating RNA Polymerase II-derived transcripts. There are potentially many fates of these transcripts. They may be translated, targeted for degradation by small RNAs, or reverse transcribed into cDNA and inserted into the genome. This section identifies several experiments aimed to detect LTR retrotransposon transcripts. Many researchers prefer to begin their analysis of retrotransposon activity with these analyses of element expression. However, one shortcoming of transcript analysis is the difficulty in assigning what transcripts originate from which locus, as there are often multiple identical copies of the same LTR retrotransposon per genome. Deep transcriptome sequencing analysis has, to some extent, resolved this problem, but the short-read nature of these sequences still means that one sequence read may match the genome dozens of times. Longer-read high-throughput sequencing technologies on the horizon will greatly facilitate the mapping of transcripts back to individual copies of LTR retrotransposons.

#### 2.2.1 Analysis of LTR Retrotransposon Transcriptional Activity

##### Analysis of Retrotransposon Transcriptional Activity by GUS Enhancer Traps

LTR retrotransposon transcriptional activity can be measured by assaying $\beta$-glucuronidase [26] protein activity from enhancer traps that have inserted into LTR retrotransposons. Enhancer trap transgenes have been spread across the *Arabidopsis* genome [26], and can be identified online (http://genetrap.cshl.edu/). Each element carries the GUS reporter coding region and a minimal promoter. When nearby transcriptional enhancers program active gene expression, the enhancer trap responds with the expression of the GUS reporter, which can be easily qualitatively and quantitatively assayed. In this case transcriptional activity of the LTR retrotransposon is inferred by the activity of retrotransposon enhancers driving GUS [20]. Using GUS enhancer traps has several benefits, including the speed in which enhancer trap lines can be obtained and assayed, and the broad range of tissues and responses to treatments that can be simultaneously assayed. Non-GUS transgene-containing plants should be used to establish a background of $\beta$-glucuronidase activity, which is present in some plant tissues.

#### 2.2.2 Analysis of Steady-State RNA Levels

##### Analysis of LTR Retrotransposon Transcript Levels by Northern Blot

LTR retrotransposon expression can be measured by extracting total or poly(A) RNA, separating the RNA based on size using electrophoresis, and blotting the RNA onto a membrane. A labeled probe complementary to the retrotransposon RNA is then hybridized to the membrane and washed. The blot is imaged to reveal bands of transcripts complementary to the labeled probe. Since LTR retrotransposons are multi-copy, polymorphic in size, and produce multiple transcripts each, bands of several sizes will be observed [17]. Northern blotting is quantitative and highly sensitive, but requires more starting RNA than reverse transcription (RT)-PCR (see below). Many investigators currently favor RT-PCR for quantitative measurement of retrotransposon transcripts. However, unlike RT-PCR, Northern blotting yields information about the size of the RNA being quantified. This may be an advantage when analyzing an LTR retrotransposon (compared to a typical gene), where multiple copies of the element are each expressing several sizes of transcripts. Northern blotting (like the RNase protection assay) is thus a powerful tool for providing a more comprehensive analysis of transcripts produced from retrotransposons.

Analysis of Retrotransposon Transcript Levels by RT-PCR

The accumulation of LTR retrotransposon transcripts can be tested using RT-PCR. In this experiment total RNA is isolated, the DNA is removed by treatment with DNase, and the RNA is reverse-transcribed *in vitro*. The resulting cDNA is then used as the template in either conventional or (q)PCR. Using different primers to generate the cDNA allows distinction of polyA+ transcripts, total RNA transcripts, as well as which strand is being transcribed. As mentioned above, PCR primers to LTR retrotransposons may be designed to amplify an individual LTR retrotransposon or whole families. For conventional PCR, the amplified DNA is subjected to gel electrophoresis, and cDNA (and thus RNA) levels are inferred by the intensity of the band. Changes in retrotransposon expression can be more finely quantified using (q)RT-PCR [21]. In this approach, the cDNA is amplified in the presence of a double-stranded DNA-binding fluorophore. As the amplification progresses, more double-stranded DNA is bound by the fluorophore and fluoresces. Initial levels of cDNA are inferred from the number of PCR cycles it takes to pass a threshold of fluorescence. For both conventional RT-PCR and (q)RT-PCR, a reference gene that is not differentially expressed in the tissue being analyzed should be examined in addition to the retrotransposon of interest, and controls lacking cDNA should be run. Values from each sample are normalized to the expression of the reference gene before comparison to other samples. Unlike Northern blotting, RT-PCR and (q)RT-PCR may be used to quickly assay the expression of many retrotransposons simultaneously and require relatively little input RNA. It is straightforward to process and analyze many samples at once using commercial qPCR instruments and the accompanying software.

### 2.2.3 Is the Retrotransposon Transcript Processed into Small RNAs?

Analysis of LTR Retrotransposon-Derived Small RNA Accumulation by Northern Blot

Endogenous small RNAs can target the transcripts of LTR retrotransposons, which leads to the degradation and post-transcriptional repression of retrotransposon activity [9]. These degraded transcripts can then be processed into small RNAs themselves, which in turn can target the retrotransposon or other transcripts for degradation, a cyclical process termed RNA interference. To determine if an LTR retrotransposon is being expressed but repressed on the post-transcriptional level by small RNAs, the retrotransposon-derived small RNAs can be detected by Northern blot [18, 27]. This assay is similar to a conventional Northern blot; however, the gel matrix is adjusted to separate small RNAs, blotting is performed using an electrical current, and the hybridization and washing conditions are adjusted [28]. Relative to RT-PCR and traditional Northern blots, large quantities of total RNA are required to visualize small RNAs, and detection of some less abundant small RNA species requires the added step of pre-purifying and enriching small RNAs from a total RNA preparation. A constitutively expressed microRNA or tRNA bands are often used as loading controls for small RNA Northern blots. In addition to the use of Northern blots to analyze LTR retrotransposon small RNAs, short-read deep sequencing has also been very successfully applied to the analysis of retrotransposon small RNAs [11, 29].

### 2.3 Are Retrotransposon Proteins Translated?

After steady-state retrotransposon RNA accumulation, the next essential step towards transposition is the synthesis of retrotransposon proteins. These include the POL and GAG proteins. The GAG protein plays a structural role in nucleocapsid formation, and POL performs the reverse transcription and integration activity necessary for new retrotransposition (Fig. 1). Although LTR retrotransposon transcripts may be present, cellular defense mechanisms can inhibit translation of the transcripts, effectively short-circuiting the retrotransposon replication cycle. Analysis of retrotransposon protein accumulation and activity is therefore necessary for determining the extent of retrotransposon activity.

#### 2.3.1 Direct Detection of LTR Retrotransposon Proteins

##### Analysis of LTR Retrotransposon Protein Levels by Western Blot

As for any protein, quantitative accumulation can be assayed by Western blot, separating out purified proteins by size using electrophoresis, transferring to a stable membrane, and using a specific antibody to detect the protein of interest. Anti-GAG antibodies have been successfully used to detect specific retrotransposon proteins in plants [30]. Another approach that has been successfully used in plants has been to purify retrotransposon GAG particles by biochemical enrichment or sucrose density gradient, followed by detection of these particles by Western blot [30, 31].

Western blots are the most direct method to determine protein accumulation. However the use of this technique is often hampered by the lack of availability of specific antibodies. Important controls include a loading control to ensure equal amounts of protein in each lane, as well as controls to ensure that the primary antibody is specific to the retrotransposon protein.

#### 2.3.2 Use of Reporter Proteins to Assay Retrotransposon Protein Accumulation

##### Gene Trap Markers of Retrotransposon Proteins

Similar to the enhancer traps described above, gene traps also use GUS or another reporter protein to monitor the activity of an LTR retrotransposon that it is inserted into. In contrast to enhancer traps, gene trap transgenes generate translational fusions between the LTR retrotransposon protein and GUS. Therefore in gene traps the GUS reporter is assaying both the endogenous LTR promoter's transcription and the translation of the retrotransposon protein [5]. Gene trap insertions in LTR retrotransposons are available for *Arabidopsis* [26] (http://genetrap.cshl.edu/), and represent a simple, rapid, and quantitative measure of retrotransposon protein accumulation. Additionally, a specific antibody is not required for this analysis. Drawbacks of this approach include the random nature of either enhancer or gene trap insertions, as well as the fact that the GUS assay measures the activity of a protein, which is an indirect measure of protein quantity.

#### Analysis of LTR Retrotransposon Activity by Reporter-Fusion Proteins

The localization and accumulation of any protein can be tested by the fusion of the protein to a fluorescent reporter, such as Green Fluorescent Protein (GFP). Custom transgenes can be designed with LTR retrotransposon proteins fused to GFP, and the spatial and temporal accumulation of that protein measured *in vivo* by fluorescent microscopy. The production of LTR retrotransposon reporter transgenes is discussed further in Subheading 2.6 (see below).

### *2.3.3 Detection of LTR Retrotransposon Protein Activity*

#### Analysis of Reverse Transcriptase Activity by PERT

Retrotransposons and retroviruses contain the coding capacity for *Reverse Transcriptase* (RT) proteins, which synthesize copied (c)DNA from an RNA template when supplied with a suitable primer (Fig. 1). Therefore the protein activity of retrotransposon and retrovirus RT can be assayed by measuring a plant's ability to reverse transcribe an exogenously provided template. In order to assay RT activity, the PERT (Product-Enhanced Reverse Transcriptase) assay was developed [32, 33]. A total protein preparation is tested for RT activity by adding an exogenous RNA (usually MS2) template with a matching DNA primer and testing for the production of reverse-transcribed cDNA by PCR. A PCR product will only be detected in the presence of RT activity in the protein preparation. PERT was originally developed to detect retroviral RT activity in mammalian cell cultures, and although results from PERT experiments in plants have not yet been published, we have successfully used PERT to examine increases in retrotransposon-derived RT activity in *Arabidopsis* mutants (Slotkin and colleagues, unpublished data). Additionally, fluorescent (F)-PERT (q)PCR and a dilution series of commercially supplied RT can be used to obtain highly quantitative measurements of RT activity from plant cell extracts [33].

In a PERT experiment, protein extracts known to lack RT activity may serve as a negative control, while protein extracts spiked with purified RT serve as a positive control. As in all PCR-based experiments, no-template-controls must be included. This consists of a PCR reaction with protein extracts where no exogenous RNA template was added. A drawback of PERT is that it does not determine what class of retrotransposon the increase in RT activity is derived from. Therefore PERT is best used in tandem with methods that directly detect the copied cDNA genome of a specific retrotransposon formed through the activity of its RT protein (described in Subheading 2.4). An advantage of PERT is its simplicity and suitability for high-throughput analyses [33].

### 2.4 Is the LTR Retrotransposon Reverse-Transcribed?

Active LTR retrotransposon transcripts are reverse-transcribed into cDNA before being integrated into the genome (Fig. 1). Since integration of the LTR retrotransposon into the plant genome may be rare, the detection of nonchromosomal copies of retrotransposon DNA represents an efficient assay to detect LTR retrotransposon activity.

#### 2.4.1 Detection of Nonchromosomal Retrotransposon Genomes by Southern Blot

Typical Southern blotting begins with the digestion of genomic DNA with a restriction enzyme (*see* Subheading 2.1.1). However, to detect nonchromosomal copies of retrotransposon DNA, Southern blotting can be performed without digestion of the genome, or with digestion with a restriction enzyme not found in the retrotransposon sequence [9, 30]. In this assay, a total DNA extract is electrophoresed, blotted onto a membrane, and hybridized with a specific labeled probe. DNA copies of LTR retrotransposons not integrated into the genome are visible on a Southern blot because they are considerably smaller than the undigested chromosomes. As with any Southern blot, prior knowledge of the probe sequence is required for this assay, and plants without active retrotransposons should be used as controls.

#### 2.4.2 Detection of Nonchromosomal Retrotransposon Genomes by Ligation-Mediated PCR

Following reverse transcription of the retrotransposon transcript, a double-stranded DNA intermediate is formed before the integration of the cDNA into the genome (Fig. 1). Using the enzyme DNA ligase, either exogenously supplied DNA adapter sequences can be ligated to the available ends of the cDNA intermediate [30], or the DNA can be self-ligated to form a circular double-stranded DNA [34]. In either case these ligated templates can be amplified by PCR that is specific for the nonchromosomal copies of the retrotransposon DNA. It is important that the input DNA is not subjected to restriction digestion before the ligation. These ligation-mediated PCR approaches are quick to perform and require little technical expertise; however, they often generate false-positive results. The difficulty with these assays may stem from the nested nature of retrotransposons in plant genomes, causing PCR primers to match many locations at the same locus, particularly after ligation of sheared chromosomal DNA. Simple controls to perform include amplifying non-ligated DNA, as well as amplifying ligated DNA from a plant without retrotransposon activity. Takeda et al. also pre-purified GAG capsids before assaying the accompanying DNA by ligation-mediated PCR, reducing the background signal derived from the chromosomal DNA [30].

### 2.5 Can New Transposition Events be Detected?

The final step in the LTR retrotransposon replication cycle is the integration of the cDNA into the plant genome. These new transposition events generate new genomic polymorphisms, which can be assayed using restriction enzymes.

#### 2.5.1 Detection of New Transpositions by Southern Blot

New insertions of LTR retrotransposons can be detected by Southern blotting of restriction-digested DNA [35, 36]. This assay differs from the Southern blots described in Subheading 2.4 above due to the use of restriction enzymes that cleave in the DNA that flanks either side of the retrotransposon. The restriction enzyme may also cut within the retrotransposon, but enough retrotransposon sequence must remain with the flanking DNA in the same DNA fragment to detect the fragment size by probing for the known sequence of the retrotransposon. The polymorphic size of the fragment, based on where the next restriction site occurs in the flanking DNA, can be detected after size separation by gel electrophoresis. A limitation of this assay is the ability to detect the polymorphic band, as the copy number of many LTR retrotransposon families is so high that individual bands cannot be resolved, or the new transposition band is masked by a preexisting retrotransposon band of the same size. Important controls for these Southern blots are the parental individuals or at least comparable inbred lines, so that a simple restriction site polymorphism between individuals is not mistaken for a new transposition event. Once the DNA flanking the LTR retrotransposon is identified (along with the retrotransposon), the putative new transposition event insertion site is cloned and sequenced to determine whether the new polymorphism is already exists in the parental genome or is a bona fide transposition event.

#### 2.5.2 Detection of New Transpositions by Transposable Element Display

Transposable element display (TED) [37] is a ligation-mediated PCR technique that, similar to the Southern blot described above, assays restriction site polymorphisms in the DNA flanking a retrotransposon element. In TED, genomic DNA is cleaved into fragments by restriction digestion, where the known end of the retrotransposon is present in a fragment with the flanking non-retrotransposon DNA. An adapter DNA fragment is ligated to the ends of all restriction fragments, and these ends are PCR-amplified using both a retrotransposon primer and the adapter primer. A nested secondary PCR reaction is performed with a labeled PCR primer, and amplicons are separated on a polyacrylamide gel [38]. As with the Southern blots described above, new transposition events appear as polymorphic bands when the individual is compared to its parent or a similar inbred line [39]. The PCR-generated DNA from these bands is isolated, cloned, and sequenced to determine where in the genome the insertion occurred [20]. This insertion site can be interrogated by sequence analysis or PCR in the parental lines to determine if this new insertion site represents the site of LTR retrotransposition. TED is extremely sensitive, due to the nature of nested PCR, and can detect transposition events that have occurred in small populations of cells within the tissue being analyzed. In contrast, Southern blot analysis is normally only able to detect transpositions that have occurred in the germline (and are thus are present in the entire plant body), or in larger somatic sectors.

### 2.6 Engineered Retrotransposition Systems

In addition to the approaches discussed here, there are other useful methods for analyzing retrotransposon and viral activity that have not yet been applied in plants. In animal cell culture systems, complex engineered retroviral vectors and transgenes have been produced to study non-LTR retrotransposons using various reporter genes. These systems include markers for expression and transposition of the retrotransposon [40]. This sensitive assay is quantitative and can be adapted for use in genetic screens designed to examine retrotransposon activity [41]. One technical difficulty precluding the design of engineered LTR retrotransposon in plants is the size of most LTR retrotransposons and the difficulty of integrating large transgenes into the plant genome. Future application of engineered retrotransposon reporter systems in plants will facilitate understanding of the factors controlling the retrotransposon replication cycle.

## 3 Conclusion and Perspectives

In contrast to mammalian genomes, where non-LTR retrotransposons have proliferated and have been extensively studied, the majority of plant genomes are composed of LTR retrotransposons. Studies of LTR retrotransposon activity and their epigenetic control continue to be at the forefront in plant systems. Recent evidence suggests different host components contribute to LTR retrotransposon silencing and likely target distinct steps in the LTR retrotransposon life cycle [9]. It is therefore important to examine the effect of various mutations and conditions on each step to ascertain the contribution of the host component to retrotransposon inhibition. The experimental steps outlined here should serve as a guide for assaying LTR retrotransposon activity and replication in plants.

## Acknowledgements

The authors gratefully acknowledge Andrea McCue and Savageethi Nuthikattu for helpful discussions and critical comments. This work was supported by the National Science Foundation grant MCB-1020499.

## References

1. Arabidopsis Genome Initiative (2000) Analysis of the genome sequence of the flowering plant *Arabidopsis thaliana*. Nature 408:796–815
2. Schnable PS et al (2009) The B73 maize genome: complexity, diversity, and dynamics. Science 326:1112–1115
3. Sharma A, Schneider KL, Presting GG (2008) Sustained retrotransposition is mediated by nucleotide deletions and interelement recombinations. Proc Natl Acad Sci U S A 105: 15470–15474
4. Bennetzen JL (2002) Mechanisms and rates of genome expansion and contraction in flowering plants. Genetica 115:29–36
5. Slotkin RK, Martienssen R (2007) Transposable elements and the epigenetic regulation of the genome. Nat Rev Genet 8:272–285
6. Lindroth AM, Cao X, Jackson JP, Zilberman D, McCallum CM, Henikoff S, Jacobsen SE (2001) Requirement of CHROMOMETHYLASE3 for maintenance of CpXpG methylation. Science 292:2077–2080
7. Zilberman D, Cao X, Jacobsen SE (2003) ARGONAUTE4 control of locus-specific siRNA accumulation and DNA and histone methylation. Science 299:716–719
8. Lippman Z, Martienssen R (2004) The role of RNA interference in heterochromatic silencing. Nature 431:364–370
9. Mirouze M, Reinders J, Bucher E, Nishimura T, Schneeberger K, Ossowski S, Cao J, Weigel D, Paszkowski J, Mathieu O (2009) Selective epigenetic control of retrotransposition in Arabidopsis. Nature 461:427–430
10. Cokus SJ, Feng S, Zhang X, Chen Z, Merriman B, Haudenschild CD, Pradhan S, Nelson SF, Pellegrini M, Jacobsen SE (2008) Shotgun bisulphite sequencing of the Arabidopsis genome reveals DNA methylation patterning. Nature 452:215–219
11. Lister R, O'Malley RC, Tonti-Filippini J, Gregory BD, Berry CC, Millar AH, Ecker JR (2008) Highly integrated single-base resolution maps of the epigenome in Arabidopsis. Cell 133:523–536
12. Lisch D, Chomet P, Freeling M (1995) Genetic characterization of the Mutator system in maize: behavior and regulation of Mu transposons in a minimal line. Genetics 139:1777–1796
13. Kato M, Miura A, Bender J, Jacobsen SE, Kakutani T (2003) Role of CG and non-CG methylation in immobilization of transposons in Arabidopsis. Curr Biol 13:421–426
14. Cao X, Jacobsen SE (2002) Locus-specific control of asymmetric and CpNpG methylation by the DRM and CMT3 methyltransferase genes. Proc Natl Acad Sci U S A 99: 16491–16498
15. Law JA, Jacobsen SE (2010) Establishing, maintaining and modifying DNA methylation patterns in plants and animals. Nat Rev Genet 11:204–220
16. Huettel B, Kanno T, Daxinger L, Bucher E, van der Winden J, Matzke AJ, Matzke M (2007) RNA-directed DNA methylation mediated by DRD1 and Pol IVb: a versatile pathway for transcriptional gene silencing in plants. Biochim Biophys Acta 1769:358–374
17. Vaillant I, Schubert I, Tourmente S, Mathieu O (2006) MOM1 mediates DNA-methylation-independent silencing of repetitive sequences in Arabidopsis. EMBO Rep 7:1273–1278
18. Lippman Z, May B, Yordan C, Singer T, Martienssen R (2003) Distinct mechanisms determine transposon inheritance and methylation via small interfering RNA and histone modification. PLoS Biol 1:e67
19. Gruntman E, Qi Y, Slotkin RK, Roeder T, Martienssen RA, Sachidanandam R (2008) Kismeth: analyzer of plant methylation states through bisulfite sequencing. BMC Bioinforma 9:371
20. Slotkin RK, Vaughn M, Borges F, Tanurdzic M, Becker JD, Feijo JA, Martienssen RA (2009) Epigenetic reprogramming and small RNA silencing of transposable elements in pollen. Cell 136:461–472
21. Teixeira FK, Heredia F, Sarazin A, Roudier F, Boccara M, Ciaudo C, Cruaud C, Poulain J, Berdasco M, Fraga MF, Voinnet O, Wincker P, Esteller M, Colot V (2009) A role for RNAi in the selective correction of DNA methylation defects. Science 323:1600–1604
22. Johnson L, Cao X, Jacobsen S (2002) Interplay between two epigenetic marks. DNA methylation and histone H3 lysine 9 methylation. Curr Biol 12:1360–1367
23. Jackson JP, Lindroth AM, Cao X, Jacobsen SE (2002) Control of CpNpG DNA methylation by the KRYPTONITE histone H3 methyltransferase. Nature 416:556–560
24. Jacob Y, Stroud H, Leblanc C, Feng S, Zhuo L, Caro E, Hassel C, Gutierrez C, Michaels SD, Jacobsen SE (2010) Regulation of heterochromatic DNA replication by histone H3 lysine 27 methyltransferases. Nature 466:987–991
25. Gendrel AV, Lippman Z, Martienssen R, Colot V (2005) Profiling histone modification patterns in plants using genomic tiling microarrays. Nat Methods 2:213–218
26. Sundaresan V, Springer P, Volpe T, Haward S, Jones JD, Dean C, Ma H, Martienssen R (1995) Patterns of gene action in plant development

revealed by enhancer trap and gene trap transposable elements. Genes Dev 9:1797–1810

27. Henderson IR, Zhang X, Lu C, Johnson L, Meyers BC, Green PJ, Jacobsen SE (2006) Dissecting Arabidopsis thaliana DICER function in small RNA processing, gene silencing and DNA methylation patterning. Nat Genet 38:721–725
28. Pall GS, Codony-Servat C, Byrne J, Ritchie L, Hamilton A (2007) Carbodiimide-mediated cross-linking of RNA to nylon membranes improves the detection of siRNA, miRNA and piRNA by northern blot. Nucleic Acids Res 35:e60
29. Kasschau KD, Fahlgren N, Chapman EJ, Sullivan CM, Cumbie JS, Givan SA, Carrington JC (2007) Genome-wide profiling and analysis of Arabidopsis siRNAs. PLoS Biol 5:e57
30. Takeda S, Sugimoto K, Kakutani T, Hirochika H (2001) Linear DNA intermediates of the Tto1 retrotransposon in Gag particles accumulated in stressed tobacco and Arabidopsis thaliana. Plant J 28:307–317
31. Bachmair A, Garber K, Takeda S, Sugimoto K, Kakutani T, Hirochika H (2004) Biochemical Analysis of Long Terminal Repeat Retrotransposons. In: Mobile Genetic Elements, Methods in Molecular Biology, eds. Miller WJ and Capy P. Humana Press 260:73–82
32. Maudru T, Peden K (1997) Elimination of background signals in a modified polymerase chain reaction-based reverse transcriptase assay. J Virol Methods 66:247–261
33. Lovatt A, Black J, Galbraith D, Doherty I, Moran MW, Shepherd AJ, Griffen A, Bailey A, Wilson N, Smith KT (1999) High throughput detection of retrovirus-associated reverse transcriptase using an improved fluorescent product enhanced reverse transcriptase assay and its comparison to conventional detection methods. J Virol Methods 82:185–200
34. Lucas H, Feuerbach F, Kunert K, Grandbastien MA, Caboche M (1995) RNA-mediated transposition of the tobacco retrotransposon Tnt1 in Arabidopsis thaliana. EMBO J 14: 2364–2373
35. Kikuchi K, Terauchi K, Wada M, Hirano HY (2003) The plant MITE mPing is mobilized in anther culture. Nature 421:167–170
36. Fukai E, Umehara Y, Sato S, Endo M, Kouchi H, Hayashi M, Stougaard J, Hirochika H (2010) Derepression of the plant Chromovirus LORE1 induces germline transposition in regenerated plants. PLoS Genet 6:e1000868
37. Lund J, Tedesco P, Duke K, Wang J, Kim SK, Johnson TE (2002) Transcriptional profile of aging in C. elegans. Curr Biol 12:1566–1573
38. Casa AM, Nagel A, Wessler SR (2004) MITE display. Methods Mol Biol 260:175–188
39. Jiang N, Bao Z, Zhang X, Hirochika H, Eddy SR, McCouch SR, Wessler SR (2003) An active DNA transposon family in rice. Nature 421:163–167
40. Rangwala SH, Kazazian HH Jr (2009) The L1 retrotransposition assay: a retrospective and toolkit. Methods 49:219–226
41. Xie Y, Rosser JM, Thompson TL, Boeke JD, An W (2011) Characterization of L1 retrotransposition with high-throughput dual-luciferase assays. Nucleic Acids Res 39:e16
42. Sabot F, Schulman AH (2006) Parasitism and the retrotransposon life cycle in plants: a hitchhiker's guide to the genome. Heredity 97: 381–388

# Chapter 14

# Detecting Epigenetic Effects of Transposable Elements in Plants


## Christian Parisod, Armel Salmon, Malika Ainouche, and Marie-Angèle Grandbastien



## Abstract

Transposable elements (TE) represent a major fraction of eukaryotic genomes and play many roles in plant epigenetics. In this chapter, we describe the use of Sequence-Specific Amplified Polymorphism (SSAP) as a reliable Transposon Display technique applicable for use in many plant species. We also discuss the interpretation of SSAP data and associated risks. This technique has potential to allow rapid screening of plant populations, especially in nonmodel or wild species.



**Key words** Transposable element, Epigenetics, SSAP, Transposition


## 1 Introduction

Transposable elements (TE) represent a major fraction of eukaryotic genomes [1] and can induce alterations in their host genome [2]. TEs indeed are highly mutagenic and silenced by overlapping epigenetic mechanisms including DNA methylation [3]. Thus, TEs represent likely candidate sequences playing a pivotal role fuelling genome structural and epigenetic reorganization [4, 5]. For a general review of their function and detection of their mobility, *see* Chapter 13. Various molecular techniques reducing genome complexity can be exploited to specifically investigate TE genome fractions [6]. Among Transposon Display strategies (i.e., high-resolution TE-anchored PCR strategy allowing the simultaneous detection of multiple insertions), Sequence-Specific Amplified Polymorphism (SSAP) is one of the most easily applicable and reliable [7, 8]. Briefly, the SSAP procedure is derived from the Amplified Fragment Length Polymorphism (AFLP) strategy, but specifically targets TEs insertions. It relies on the amplification of digested genomic DNA with primers designed at the border of TEs and generates a pool of labeled fragments containing the


Charles Spillane and Peter C. McKeown (eds.), *Plant Epigenetics and Epigenomics: Methods and Protocols*, Methods in Molecular Biology, vol. 1112, DOI 10.1007/978-1-62703-773-0_14,

termini of inserted copies of a given TE and its flanking genomic region [9]. SSAP usually generates highly polymorphic markers that allow to reliably assessing patterns of genetic diversity within and among groups [10–12]. SSAP polymorphism may result from molecular changes at insertion sites that modify the size of the amplification product [13], but comparative SSAP banding patterns can also offer reliable insights on the genome dynamics of TE fraction among related lineages. Relying on proper TE-specific primers represents the decisive step for implementing a reliable SSAP. The reader willing to design TE-specific primers for an SSAP procedure or any of its derivatives allowing to reliably amplify genomic regions flanking insertions of a given TE would profitably consult the literature on that topic [8, 14].

In this chapter, we present a recent modification of the SSAP protocol using restriction enzymes with differential sensitivity to DNA methylation at the digestion step. This methyl-sensitive derivative of SSAP has been named Methyl-Sensitive Transposon Display (MSTD; Fig. 1a) and provides useful knowledge on the methylation environment of TE insertions [15, 16]. The isoschizomers MspI and HpaII are widely used for methyl-sensitive displays (e.g., ref. [17–19]), as both enzymes recognize the same tetranucleotide sequence (5′-CCGG-3′), but present different sensitivities to DNA methylation (ref. [20]; Fig. 1b). HpaII is sensitive to methylation of any cytosine at both strands (5′-CCGG-3′), whereas MspI cuts methylated internal cytosine (5′-C5mCGG-3′). These properties allow assessing the methylation status of internal cytosine at restriction sites (CpG methylation). However, MspI is sensitive to methylation of the external cytosine (5′-5mCCGG-3′). Hence, methylation of external cytosine on both strands (CpCpG methylation: 5′-5mCCGG-3′ and 5′-5mC5mCGG-3′) may not produce bands with this MSTD. Since HpaII cleaves when the external cytosine is methylated on one strand, whereas MspI does not, hemimethylated CpCpG sites can be detected with this MSTD.

In addition to the limits inherent to the SSAP, MSTD profiles have to be interpreted with caution because CpCpG methylation on both strands prevents the enzymes from cutting and the technique is thus blind to heavily methylated portions of the genome. The absence of selected bands in specific samples might thus reveal either restructuring of the TE insertion or increased methylation in the vicinity of the TE insertion. Accordingly, a band specific from selected samples might correspond to a transposition event or demethylation in the vicinity of a TE insertion. MSTD data are biased toward non-heavily methylated regions and chiefly assess methylation changes in this particular fraction of the genome. Furthermore, it should always be kept in mind that MSTD offers insights about the methylation status of sequences flanking a particular TE insertion (i.e., the CCGG site next to a TE insertion) and does not necessarily reflects methylation changes affecting the TE insertion itself.

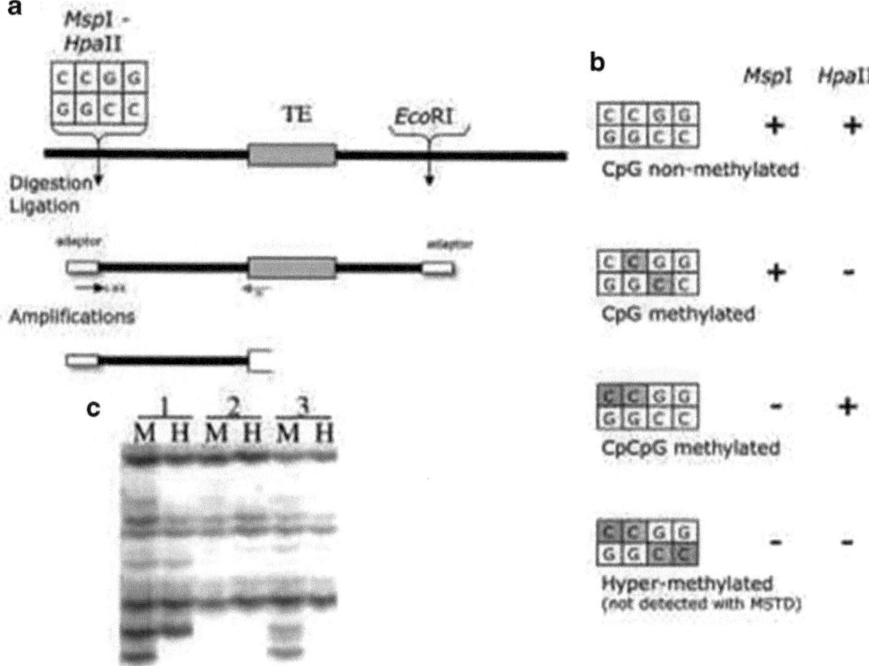


**Fig. 1** Principle of the methyl-sensitive transposon display (MSTD). (**a**) Schematic representation of the high-resolution TE-anchored PCR strategy allowing the simultaneous detection of multiple insertions. After digestion of genomic DNA with rare cutter (e.g., MspI/HpaII) and frequent cutter (e.g., EcoRI) restriction enzymes, adaptors are ligated to DNA fragments. PCR amplifications are carried out using a primer complementary to the rare cutter adaptor and a labeled (*asterisk*) primer specific to the targeted transposable element (TE). (**b**) Methylation sensitivity of isoschizomer enzymes (MspI and HpaII) and interpretation of resulting banding pattern as a function of presence (+)/absence (–) of a given MSTD band. (**c**) An example of MSTD banding pattern for three samples (1–3). Comparison of band presence/absence in MspI (M) and HpaII (H) profiles reveals the methylation state of restriction sites flanking the corresponding TE insertion

MSTD was successfully employed to detect methylation changes following genomic shocks such as interspecific hybridization and/or genome duplication (15). Insights offered by this method can be valuably compared to data derived from Methyl-Sensitive Amplified Polymorphism (MSAP), combining AFLP and methylation-sensitive restriction enzymes to allow random detection of methylation changes across the genome [21]. When contrasting these two methods, Parisod et al. [15] were able to detect significantly more CpG methylation changes in regions flanking TE insertions than in random sequences following recent hybridization and genome duplication in Spartina, indicating that TEs were the most (epigenetically) targeted compartment subject to rapid evolution during allopolyploid speciation. We anticipate that the MSTD method will remain useful for rapid screening of populations, most particularly in nonmodel or wild species where genomic resources and information are limited.

## 2 Materials

### 2.1 Digestion

1. Tango Buffer [10×]: 33 mM Tris-acetate (pH 7.9); 10 mM Mg-acetate; 66 mM K-acetate; and 0.1 mg/mL BSA.
2. EcoRI [10 U/mL]: rare cutter enzyme (5′-GAATTC-3′) (*see* **Note 1**).
3. MspI/HpaII [10 U/mL]: frequent cutter enzymes recognizing the same tetranucleotide sequence (5′-CCGG-3′), but displaying differential sensitivity to DNA methylation.

### 2.2 Ligation

1. EcoRI-adaptors [100 mM]: 5′-CTCGTAGACTGCGTACC-3′ and 5′AATTGGTACGCAGTCTAC-3′. Preparation: mix equal volumes of the two adaptors [final concentration: 50 mM] and warm up to 95 °C for 5 min, then allow to cool down to room temperature. Then dilute at 1/10 for a final concentration of 5 mM. MspI/HpaII-adaptors [100 mM]: 5′-GACGATGAGTCTAGAA-3′ and 5′-CGTTCTAGACTCATC-3′. Preparation: mix equal volumes of the two adaptors [final concentration: 50 mM] and warm up to 95 °C for 5 min, then allow to cool down to room temperature.
2. ATP [20 mM].
3. T4 DNA ligase [5 U/mL].

### 2.3 PCR Preselective Amplification

1. EcoRI + A primer [10 mM]: 5′-GACTGCGTACCAATTCA-3′.
2. MspI/HpaII + C primer [10 mM]: 5′-GATGAGTCTAGAA CGGC-3′.
3. Rxn Buffer [10×]: 200 mM Tris pH 8.4 + 500 mM KCl.
4. Equimolar dNTPs [10 mM].
5. $MgCl_2$ [25 mM].
6. Taq polymerase [5 U/mL].

### 2.4 PCR Preselective Amplification

1. Labeled TE-specific primers (*see* **Note 2**).
2. MspI/HpaII selective primers were similar to that of preselective primer, with the addition of two variable nucleotides (=MspI/HpaII + CXX primer).
3. Primers used for this step were otherwise similar to those listed in Subheading 2.3.

## 3 Methods

### 3.1 Digestion (See Note 3)

1. Add 5 mL of Tango Buffer to 12.7 mL of sterile water.
2. Add 0.1 mL (1 U) of EcoRI.

3. Add 0.2 mL (2 U) of MspI (alternatively, HpaII) and gently mix.
4. Add 250 ng of DNA in 7 mL to this mix (final volume 25 mL) and gently mix.
5. Incubate at 37 °C for 3 h.
6. Deactivate restriction enzymes at 70 °C for 15 min.

### 3.2 Ligation (See Note 3)

1. Add 3 mL of Tango Buffer to 8.5 mL of sterile water.
2. Add 1 mL of ATP.
3. Add 1 mL of EcoRI-adaptors.
4. Add 1 mL of MspI/HpaII-adaptors and vortex.
5. Add 0.5 mL (5 U) of T4 DNA ligase and gently mix.
6. Add this ligation mix (15 mL) to the 25 mL of digestion mix (final volume 40 mL) and gently mix.
7. Incubate at room temperature (22 °C) overnight.
8. (Optional) 5 μL of product can be visualized by electrophoresis on a 1 % agarose gel and stained with ethidium bromide or similar in order to verify the success of digestion.
9. Dilute the digestion–ligation mix four times (e.g., 30 μL of product in 90 μL of sterile water) (*see* **Note 4**).

### 3.3 Performance of PCR Preselective Amplification

1. Add 2 μL of reaction (Rxn) Buffer to 12.7 μL of sterile water.
2. Add 0.5 μL of dNTP.
3. Add 1.6 μL of $MgCl_2$.
4. Add 0.5 μL of EcoRI + primer A.
5. Add 0.5 μL of MspI/HpaII + primer C and vortex.
6. Add 0.2 μL of Taq polymerase (1 U).
7. Add 18 μL of this preselective mix to 2 μL of diluted digestion–ligation mix (to a final volume 20 μL).
8. Place in a thermocycler to perform this PCR amplification: 94 °C for 180 s, followed by 28 cycles at 94 °C for 30 s, 60 °C for 60 s, and 72 °C for 60 s, and a final extension at 72 °C for 180 s.
9. Dilute the preselective amplification products 1:10 with sterile water (e.g., 10 μL of product in 190 μL of water).

### 3.4 Performance of PCR Selective Amplification

1. Add 2 μL of Rxn Buffer to 11.1 μL of sterile water.
2. Add 0.5 μL of dNTP.
3. Add 1.6 μL of $MgCl_2$.
4. Add 0.8 μL of TE-specific primer.
5. Add 0.8 μL of MspI/HpaII + CXX primer and vortex.

6. Add 0.2 μL of Taq polymerase (1 U).
7. Add 17 μL of this selective mix to 3 μL of diluted preselective amplification product (final volume 20 μL).
8. Place in a thermocycler to perform this touch-down PCR amplification: 94 °C for 120 s, followed by 13 cycles at 94 °C for 30 s, 65 °C to 56 °C (decreasing by 0.7 °C per cycle) for 30 s, and 72 °C for 60 s, followed by 25 cycles at 94 °C for 30 s, 56 °C for 30 s, and 72 °C for 60 s, and a final extension at 72 °C for 300 s.
9. Prepare the amplification products according to your electrophoresis protocol (*see* **Note 4**).

## 4 Notes

1. EcoRI is very widely used but can be variable in its sensitivity to CpG methylation [20]. SSAP protocols using Csp6 instead of EcoRI have been developed [8, 14] and might be profitably used for MSTD. Unfortunately, Csp6 is a four-base restriction enzyme (i.e., frequent cutter) and might provide too much SSAP bands for the analysis of complex genomes. Furthermore, with such frequent cutter, it might happen that the TE itself presents a restriction site. This would induce the amplification of a band internal to the TE instead of a band containing the flanking genomic DNA, resulting in confusing results, especially in the case of retrotransposons bordered by two identical long terminal repeats.
2. It is vital to label the TE primer in order to highlight bands containing the termini of an inserted TE and its flanking genomic region. The TE primer can be radioactively labeled with P33 or with fluorochromes. Amplification products labeled with P33 can be visualized after electrophoresis on 6 % Long Ranger denaturing gel for 5 h (75 V, limited to 2,000 W) by autoradiography. Amplification products labeled with fluorochromes can be visualized with automatic sequencers after electrophoresis.
3. MspI and HpaII have to be used on the same samples in parallel in order to provide an MSTD. Accordingly, preparing two mixes in parallel at each step (one for the MspI reactions and one for the HpaII reactions) can improve reliability and comparability of the MSTD profiles. Although SSAP and MSTD approaches generate reliable consistent patterns, it is strongly advisable to perform the protocol several times on selected samples in order to estimate the error rate (*see* refs. 22, 23 for further details).
4. Diluted digestion–ligation product can be stored at –20 °C.

## References


1. Gaut BS, Ross-Ibarra J (2008) Selection on major components of angiosperm genomes. Science 320:484–486
2. McClintock B (1984) The significance of responses of the genome to challenge. Science 16:792–801
3. Kato M, Miura A, Bender J, Jacobsen SE, Kakutani T (2003) Role of CG and non-CG methylation in immobilization of transposons in Arabidopsis. Curr Biol 13:421–426
4. Michalak P (2009) Epigenetic, transposon and small RNA determinants of hybrid dysfunctions. Heredity 102:45–50
5. Parisod C, Alix K, Just J, Petit M, Sarilar V, Mhiri C, Ainouche M, Chalhoub B, Grandbastien MA (2010) Impact of transposable elements on the organization and function of allopolyploid genomes. New Phytol 186:37–45
6. Kalendar R, Flavell A, Ellis TH, Sjakste T, Moisy C, Schulman AH (2011) Analysis of plant diversity with retrotransposon-based molecular markers. Heredity 106:520–530
7. Waugh R, McLean K, Flavell AJ, Pearce SR, Kumar A, Thomas BBT, Powell W (1997) Genetic distribution of Bare-1-like retrotransposable elements in the barley genome revealed by sequence-specific amplification polymorphisms (SSAP). Mol Gen Genet 253:687–694
8. Syed NH, Flavell AJ (2006) Sequence-specific amplification polymorphisms (SSAPs): a multi-locus approach for analyzing transposon insertions. Nat Protoc 1:2746–2752
9. Melayah D, Lim KY, Bonnivard E, Chalhoub B, De Borne FD, Mhiri C, Leitch AR, Grandbastien MA (2004) Distribution of the Tnt1 retrotransposon family in the amphidiploid tobacco (*Nicotiana tabacum*) and its wild Nicotiana relatives. Biol J Linn Soc 82: 639–649
10. Petit M, Lim KY, Julio E, Poncet C, de Borne FD, Kovarik A, Leitch AR, Grandbastien MA, Mhiri C (2007) Differential impact of retrotransposon populations on the genome of allotetraploid tobacco (*Nicotiana tabacum*). Mol Gen Genet 278:1–15
11. Sanz AM, Gonzalez SG, Syed NH, Suso MJ, Saldana CC, Flavell AJ (2007) Genetic diversity analysis in Vicia species using retrotransposon-based SSAP markers. Mol Gen Genet 278:433–441
12. Tam SM, Mhiri C, Vogelaar A, Kerkveld M, Pearce SR, Grandbastien MA (2005) Comparative analyses of genetic diversities within tomato and pepper collections detected by retrotransposon-based SSAP, AFLP and SSR. Theor Appl Genet 110:819–831
13. Petit M, Guidat C, Daniel J, Denis E, Montoriol E, Bui QT, Lim KY, Kovarik A, Leitch AR, Grandbastien MA, Mhiri C (2010) Mobilization of retrotransposons in synthetic allotetraploid tobacco. New Phytol 186:135–147
14. Tam S, Mhiri C, Grandbastien M-A (2006) Transposable elements and the analysis of plant biodiversity. In: Morot-Gaudry J, Lea P, Briat J (eds) Functional plant genomics. Sciences Publishers, Enfield, NH, pp 529–558
15. Parisod C, Salmon A, Zerjal T, Tenaillon M, Grandbastien MA, Ainouche ML (2009) Rapid structural and epigenetic reorganization near transposable elements in hybrid and allopolyploid genomes in Spartina. New Phytol 184:1003–1015
16. Kashkush K, Khasdan V (2007) Large-scale survey of cytosine methylation of retrotransposons and the impact of readout transcription from long terminal repeats on expression of adjacent rice genes. Genetics 177:1975–1985
17. Cervera MT, Ruiz-Garcia L, Martinez-Zapater JM (2002) Analysis of DNA methylation in *Arabidopsis thaliana* based on methylation-sensitive AFLP markers. Mol Genet Genomics 268:543–552
18. Shaked H, Kashkush K, Ozkan H, Feldman M, Levy AA (2001) Sequence elimination and cytosine methylation are rapid and reproducible responses of the genome to wide hybridization and allopolyploidy in wheat. Plant Cell 13:1749–1759
19. Takata M, Kiyohara A, Takasu A, Kishima Y, Ohtsubo H, Sano Y (2007) Rice transposable elements are characterized by various methylation environments in the genome. BMC Genomics 8:469
20. Roberts RJ, Vincze T, Posfai J, Macelis D (2010) REBASE—a database for DNA restriction and modification: enzymes, genes and genomes. Nucleic Acids Res 38:D234–D236
21. Salmon A, Ainouche ML, Wendel JF (2005) Genetic and epigenetic consequences of recent hybridization and polyploidy in Spartina (Poaceae). Mol Ecol 14:1163–1175
22. Bonin A, Bellemain E, Eidesen PB, Pompanon F, Brochmann C, Taberlet P (2004) How to track and assess genotyping errors in population genetics studies. Mol Ecol 13:3261–3273
23. Pompanon F, Bonin A, Bellemain E, Taberlet P (2005) Genotyping errors: causes, consequences and solutions. Nat Rev Genet 6: 847–859

Chapter 15

# Detection and Investigation of Transitive Gene Silencing in Plants


Leen Vermeersch, Nancy De Winne, and Ann Depicker



## Abstract

RNA-induced post-transcriptional silencing is a common tool in functional gene analysis and its application in crop improvement is widely investigated. However, its specificity might be impaired by off-target silencing as a result of transitivity. Generally transitivity is investigated by the detection of secondary siRNAs; however, these tests fail to demonstrate the siRNA's bioactivity. Here, we describe protocols to detect the occurrence of transitive silencing across an endogene by using a reporter *GUS* gene. In addition, we provide a setup to test the influence of a sequence of interest present in the primary target on the progression of transitivity.



**Key words** Post-transcriptional gene silencing (PTGS), Transitive silencing, siRNA, β-Glucuronidase (GUS), Southern blot, *Arabidopsis thaliana*


## 1 Introduction

RNA interference (RNAi) denotes both transcriptional as well as post-transcriptional silencing induced by small RNAs (smRNAs) [1]. In all RNA-induced silencing systems, smRNAs arise from the processing of double-stranded RNA (dsRNA) into 21–24 nucleotide (nt) smRNAs by an RNaseIII-like ribonuclease, known in plants as DICER-LIKE (DCL). Incorporation of these smRNAs into ARGONAUTE (AGO) containing effector complexes channels sequence-specific degradation and translational repression to the target RNA or guides cytosine methylation of homologous DNA sequences.

RNAi is often exploited to analyze gene functions, but has also become a commercially focused application that covers a wide spectrum of potentialities, from protection of plants against pathogen attack to fine-tuning of metabolic pathways. Silencing inducers consist either of modified viruses, sense or antisense gene constructs, hairpin constructs, or artificial microRNAs (amiRNAs) producing dsRNA designed to specifically target a gene of interest.


Charles Spillane and Peter C. McKeown (eds.), *Plant Epigenetics and Epigenomics: Methods and Protocols*, Methods in Molecular Biology, vol. 1112, DOI 10.1007/978-1-62703-773-0_15, © Springer Science+Business Media New York 2014

Nevertheless, in nematodes, fungi, and plants, amplification of the original silencing trigger-derived smRNA pool occurs, enabling the occurrence of off-target silencing [2–7]. This amplification involves the RNA-dependent RNA polymerase (RDR)-mediated dsRNA synthesis from target gene transcripts and conversion of this dsRNA by DCL to secondary small interfering RNAs (siRNAs). When these secondary siRNAs are produced from sequences non-homologous to the silencing inducer sequence, this can lead to transitive silencing of a secondary target that is not directly targeted by the silencing trigger [5, 6]. Depending on the setup and the primary target genes, spreading is observed in both directions, only in the 5′–3′ direction [8, 9] or not at all. Strikingly, transitivity is rarely observed when silencing is targeted to endogenes [9–12].

In plants, most of the working mechanism of transitivity was uncovered when silencing was targeted against a transgene by a silencing inducer containing only part of the primary target. To demonstrate transitivity, the presence of secondary siRNAs from sequences neighboring the primary target sequence is investigated. The latter can be done by directly analyzing the siRNA content through northern blotting [3]. However, the detection of secondary siRNAs does not provide information about their functionality. To validate the functionality of secondary siRNAs, an XYZ transitive silencing assay can be performed, for which two target RNAs are introduced: a primary target (Y) with partial homology to the silencing inducing dsRNA (X), and a secondary target (Z) nonhomologous to the trigger X, but containing homology to the remaining part of the primary target Y (Fig. 1a).

To easily assess the silencing of the primary and, if applicable, the secondary target, the latter generally encodes a reporter gene like *green fluorescent protein* (*GFP*) or *β-glucuronidase* (*GUS*). The use of *GFP* as a target offers the advantage of noninvasive and simple visualization of silenced and non-silenced cells. On the other hand, the *GUS* reporter gene allows an easy determination of the protein concentration, reflecting the degree of silencing. Moreover, as post-transcriptional gene silencing (PTGS) reduces expression not only by cleavage of the target mRNA, but also through translational repression, it is worthwhile to focus on protein levels in addition to mRNA expression levels of the target gene. The relevance of translational repression was demonstrated in two reports on *in trans* and transitive post-transcriptional silencing, which revealed an up to eightfold greater reduction at the protein level compared to the mRNA level [13, 14].

Here, we describe two XYZ transitive silencing assays in *Arabidopsis thaliana* that are easily adjustable to analyze several aspects of transitive silencing. The first configuration (Fig. 1b) is intended to investigate if *in trans* silencing of an endogene triggers the formation of endogenous functional secondary siRNAs enabling silencing of a secondary target consisting of a translational

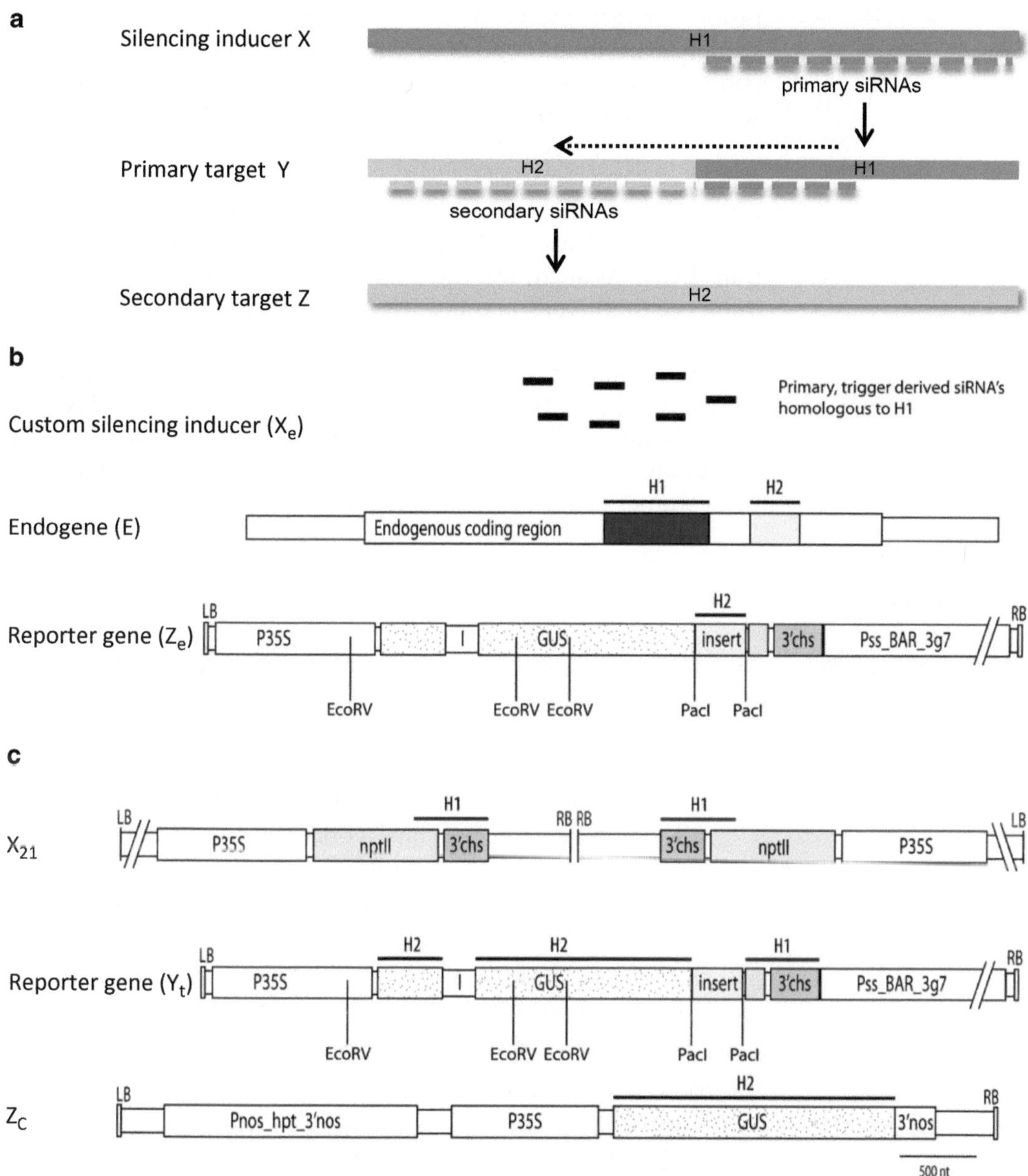


**Fig. 1** Schematic outline of the XYZ transitive silencing assay. (**a**) In the XYZ transitive silencing assay, the silencing inducer X and the primary target Y are partially homologous (indicated as H1). The primary target Y is also partially homologous to the secondary target Z (indicated as H2). (**b**) Transitive silencing configuration to detect transitivity across an endogene. In this case the endogene is the primary target and the *GUS* reporter gene is the secondary target. (**c**) Transitive silencing configuration to test the influence of a sequence of interest (insert) on transitivity progression. Here, the *GUS* reporter gene is used as a primary target. *P35S* cauliflower mosaic virus 35S promoter, *GUS β-glucuronidase*, *3′chs* 3′ UTR of the *chs* gene of snapdragon, *Pss* promoter of the small subunit of *ribulose-1,5-biphosphate carboxylase*, *BAR bialophos acetyltransferase*-coding sequence conferring phosphinothricin resistance, *3g7* 3′ UTR of the *Agrobacterium tumefaciens octopine T-DNA* gene 7, *nptII neomycin phosphotransferase gene*, *Pnos nopaline synthase* promoter, *hpt hygromycin phosphotransferase*-coding sequence, *3′nos* 3′ UTR of the *nopaline synthase* gene, *RB* right T-DNA border, *LB* left T-DNA border

fusion between the *GUS* reporter gene and part of the endogene sequence nonhomologous to the silencing inducer. The second configuration (Fig. 1c) was designed to examine the influence of several sequences on the spreading of silencing by inserting in the primary target a sequence of interest between the primary target region and the reporter *GUS* region homologous to the secondary target. This configuration was previously exploited to examine the differential behavior of transgenes and endogenes in supporting transitivity. As such, the distance transitivity can bridge [15], the influence of introns on transitivity processivity [16] and the transitive silencing of endogenes [13] was studied. The use of the *GUS* reporter gene allows an easy assessment of transitive silencing at the protein level. To further investigate transitivity, the siRNA and RNA quantity and the methylation level of the corresponding DNA could be determined (good protocols are described in previous volumes of *Methods in Molecular Biology*).

## 2 Materials

### 2.1 A. thaliana Floral Dip Transformation

1. *Agrobacterium tumefaciens* C58C1Rif$^R$(pMP90, Gm$^R$) harboring the appropriate binary T-DNA vector.
2. Plates containing selective YEB medium: 5 g/L beef extract, 1 g/L yeast extract, 5 g/L peptone, 5 g/L sucrose, 2 mM $MgSO_4$, 1.5 % agar. The C58C1Rif$^R$ (pMP90, Gm$^R$) strain is resistant to rifampicin (final concentration 100 mg/L) and gentamicin (final concentration 40 mg/L). The T-DNA vector with the reporter gene ($Y_t$ or $Z_e$) confers spectinomycin resistance (final concentration 100 mg/L).
3. *Arabidopsis thaliana* (L.) Heynh. plants, ecotype Columbia (Col-0) with the appropriate transgenic background (*see* **Note 1**).
4. Aracon bases and tubes (Lehle Seeds, Round Rock, TX).
5. Saran wrap (Dow Chemical, Midland, MI).
6. Luria Bertani (LB$^-$) medium: 10 g/L tryptone, 5 g/L yeast extract, 10 g/L NaCl in $ddH_2O$.
7. Dipping solution: 10 % (w/v) sucrose, 0.05 % (v/v) Silwet L77 (Lehle Seeds) in $ddH_2O$ (*see* **Note 2**).

### 2.2 Selection of Transgenic Lines and Identification of Single-Locus Lines

1. Selective K1 growth medium: 4.3 g/L Murashige and Skoog (MS) salts (Gibco BRL, Gaithersburg, MD), 0.5 g/L MES (2-(*N*-Morpholino) ethanesulfonic acid monohydrate, Duchefa, Haarlem, The Netherlands) and 10 g/L sucrose in $ddH_2O$. Set to pH 5.7 with 1 M KOH. Add 7 g/L agar and sterilize by autoclaving at 1 bar over-pressure (121 °C) for 20 min. Cool down to approximately 60 °C. In a flow bench, add the selective agents (*see* Subheading 2.2, **items 2, 3**, and **4**;

*see* **Note 3**) and pour 80–100 mL in round Petri dishes (150 × 25 mm). To prevent condensation, allow the agar to set before closure of the plates. Store the plates in a sealed sterile bag and keep at room temperature for up to 2 months.

2. To select for the transgenes, add either 50 mg/L kanamycin A sulfate (500 μL 100 mg/mL stock solution in water (Duchefa, Haarlem, The Netherlands), single use aliquots), 10 mg/L phosphinothricin (PPT) (100 μL 100 mg/mL stock solution in $ddH_2O$, Sigma-Aldrich, Saint Louis, MO 63103, USA) or 20 mg/L hygromycin B (400 μL 50 mg/mL stock solution in $ddH_2O$, Duchefa, Haarlem, The Netherlands), or a combination of these (*see* **Note 4**).
3. As fungicidal agent 50 mg/L nystatin (stock solution of 50 mg/mL in dimethylsulfoxide [DMSO], Duchefa, Haarlem, The Netherlands) is added (*see* **Note 4**).
4. When seeds derived from a floral dip are used (T1 seed stock), 500 mg/mL carbenicillin disodium (1 mL 500 mg/mL stock solution in $ddH_2O$, Duchefa, Haarlem, The Netherlands) is added to select against *A. tumefaciens* (*see* **Note 5**).
5. Miracloth (Merck Chemicals, Nottingham, UK).
6. 75 % EtOH.
7. 5 % bleach with 0.1 % Tween 20: 42 mL 12 % NaOCl, 100 μL Tween 20 (polyoxyethylene sorbitan monolaurate, Sigma-Aldrich, Saint Louis, MO 63103, USA), 58 mL sterile $ddH_2O$ (*see* **Note 6**).
8. 0.1 % agarose in sterile $ddH_2O$ (sterilized by autoclaving).

### 2.3 Selection of Single Copy Lines Containing $Y_t$ or $Z_e$

#### 2.3.1 Creating a Digoxigenin-Labeled Probe

1. Non-DIG template for the probe created by standard PCR (*see* **Note 7**).
2. Taq DNA polymerase 5 U/μL and 10× PCR buffer (Invitrogen, Carlsbad, USA).
3. 50 mM $MgCl_2$.
4. Digoxigenin (DIG) dNTPs (PCR DIG-probe synthesis kit, Roche Applied Science, Mannheim, Germany).
5. Forward and reverse primers used to create the non-DIG template at 100 μM.
6. High pure PCR product purification kit (Roche Applied Science, Mannheim, Germany).

#### 2.3.2 Blotting DNA on a Membrane

1. 5 μg EcoRV-digested DNA of the selected single locus lines (*see* **Note 8**).
2. 3 μL Smartladder (Eurogentec, Seraing, Belgium).
3. 0.2 ng EcoRV-digested $Y_t$ or $Z_e$ plasmid DNA (*see* **Note 9**).

4. 10× DNA loading dye: 0.25 % bromophenol blue, 0.25 % xylene cyanol, 15 % ficoll in $ddH_2O$.
5. Technical 10 % SDS (sodium dodecyl sulfate).
6. Horizontal gel electrophoresis system for gels of about 15×25 cm.
7. 10× TAE: 0.4 M Tris, 0.2 M acetic acid, 10 mM EDTA.
8. 1.5 % agarose, 1× TAE gel, and 1× TAE buffer.
9. Ethidium bromide (EtBr).
10. Two glass dishes and two glass plates.
11. Whatman paper.
12. Amersham Hybond $N^+$ filter (GE Healthcare limited, Buckinghamshire, England).
13. Saran wrap (Dow Chemical, Midland, MI).
14. A staple of 15 cm paper napkins.
15. 0.25 M HCl.
16. Denaturation buffer (pH 14): 1.5 M NaCl, 0.5 M NaOH.
17. Neutralization buffer (set to pH 7.5 with HCl): 1.5 M NaCl, 0.5 M Tris.
18. 20× SSC (saline sodium citrate): 0.3 M $Na_3$ citrate, 3 M NaCl.
19. Biorad GS genelinker UV chamber (Biorad, Hercules, CA).

#### 2.3.3 Hybridization and Detection

1. Hybridization oven and tubes.
2. DIG Easy Hyb buffer (Roche Applied Science, Mannheim, Germany).
3. 1 L 2× SSC/0.1 % SDS: dissolve 1 g SDS in 1 L 2× SSC (autoclave).
4. 1 L 1× SSC/0.1 % SDS: dissolve 1 g SDS in 1 L 1× SSC (autoclave).
5. Buffer 1: 100 mM maleic acid (58.05 g/5 L), 150 mM NaCl (43.85 g/5 L), bring to approximately pH 7.5 with 40 g NaOH. Check the pH and adjust with NaOH and HCl.
6. Wash buffer: buffer 1 containing 0.3 % Tween20 (autoclave).
7. Block buffer: buffer 1 containing 1 % blocking powder (dissolves during autoclaving, Roche Applied Science, Mannheim, Germany). Store at 4 °C.
8. B3 buffer: 100 mM Tris/HCl pH 9.5, 100 mM NaCl. Make a 10× stock (1 M) of the components and mix them (100 mL of each + 800 mL $ddH_2O$) just before use.
9. Anti-DIG-AP (Roche Applied Science, Mannheim, Germany).
10. 25 mM CDP-Star (100×, Roche Applied Science, Mannheim, Germany).

11. EtOH and Dettol.
12. Freezer bags and tape.
13. Whatman paper.
14. Cassette and film to expose.

*2.3.4 Strip the Membrane*

1. 0.2 M NaOH/0.1 % SDS.
2. 2× SSC.

***2.4 Crossing of Lines Expressing the Silencing Inducer with Lines Expressing the Primary and Secondary Target***

1. Aracon bases and tubes (Lehle Seeds, Round Rock, TX).
2. Sewing thread or other fine labeling.

***2.5 Protein Extraction from A. thaliana Seeds***

1. Grinding ball mill device MM200 (Retsch, Haan, Germany).
2. Cooled (4 °C) centrifuge 5417R (Eppendorf, Hamburg, Germany).
3. Metal balls of 4 mm diameter.
4. 0.1 M phosphate buffer (pH 7): for 250 mL (*see* **Note 10**):
   (a) 49 mL 0.2 M $NaH_2PO_4$.
   (b) 76 mL 0.2 M $Na_2HPO_4 \cdot 12H_2O$.
   (c) 125 mL sterile $ddH_2O$.
5. GUS extraction buffer (GUS EB 2×): for 50 mL:
   (a) 1 mL 0.5 M EDTA (pH 8) (final concentration 10 mM).
   (b) 50 μL Triton X-100 (final concentration 0.1 %, Sigma-Aldrich, Saint Louis, MO 63103, USA).
   (c) 25 mL 0.1 M phosphate buffer (final concentration 0.05 M) (*see* Subheading 2.5, **item 4**).
   (d) 24 mL sterile $ddH_2O$.
   (e) 35 μL β-mercaptoethanol (final concentration 10 mM, Sigma-Aldrich, Saint Louis, MO 63103, USA) (add in the fume hood).
6. 100 % (v/v) glycerol.

***2.6 Determination of the Protein Concentration in Leaf Extracts***

1. Transparent 96-well microtiter plates (flat bottom, untreated, polystyrene; Nunc, Roskilde, Denmark).
2. Microtiter plate stickers.
3. VERSAmax tunable microplate reader (Molecular Devices, Sunnyvale, CA).
4. Softmax® Pro Software 3.0 (Molecular Devices, Sunnyvale, CA).

5. BIO-RAD Protein assay reagents (BIO-RAD, Hercules, CA). Add 10 mL dye to 35 mL sterile $ddH_2O$ in the fume hood, filter to avoid precipitate, store at 4 °C in the dark.
6. GUS EB 2× 1/5. Dilute 5 mL GUS EB 2× (*see* Subheading 2.5, **item 5**) with 20 mL sterile $ddH_2O$.
7. 0.1 mg/mL bovine serum albumin (BSA) (Sigma-Aldrich, Saint Louis, MO 63103, USA) stock solution in sterile $ddH_2O$ (stored in single-use aliquots at −20 °C).

#### 2.7 Fluorimetric Measurement of the GUS Activity in Leaf Extracts

1. Black polysorp 96 well plates (Nunc, Roskilde, Denmark).
2. FLUOstar OPTIMA fluorescence plate reader with FLUOstar optima software (BMG LABTECH GmbH, Ortenberg, Germany).
3. GUS EB 2× (*see* Subheading 2.5, **item 5**).
4. 10 mM 4-Methylumbelliferone (4-MU, Sigma-Aldrich, Saint Louis, MO 63103, USA): dissolve 0.01982 g 4-MU in 10 mL sterile $ddH_2O$ (*see* **Note 11**).
5. 1 mM 4-MU: add 100 μL 10 mM 4-MU to 900 μL sterile $ddH_2O$ (*see* **Note 11**).
6. 4 mM 4-Methylumbelliferyl-β-D-glucuronide hydrate (4-MUG, Sigma-Aldrich, Saint Louis, MO 63103, USA) in GUS EB 2× (*see* **Note 12**).
7. 72 U/μL β-glucuronidase (GUS) enzyme, aqueous glycerol solution (Sigma-Aldrich, Saint Louis, MO 63103, USA).

## 3 Methods

#### 3.1 Experimental Design for Investigation of Transitivity Along an Endogene

In this configuration the endogene of interest (E) acts as primary target, which is silenced by a custom designed silencing inducer $X_e$. The transgenic reporter construct will be used as secondary target, $Z_e$ (Fig. 1b).

1. The silencing inducer should be customized to your experimental question and designed to specifically silence your endogene of interest. Silencing might be obtained by different transgenic silencing loci creating, for instance, long dsRNAs, hairpin RNAs, or amiRNAs. Cloned sequence repertoires exist with gene-specific tags cloned either in Gateway entry vectors or hairpin constructs [17]. Notice that the silencing inducer is not allowed to show homology with the secondary target, except for the promoter, if not inevitable.
2. To construct the secondary target $Z_e$, the endogenous sequence from which one wants to test the existence of siRNAs has to be cloned into the PacI restriction site of the P35S_GUS_PacI_npt3'chs reporter gene. Use an endogene sequence upstream

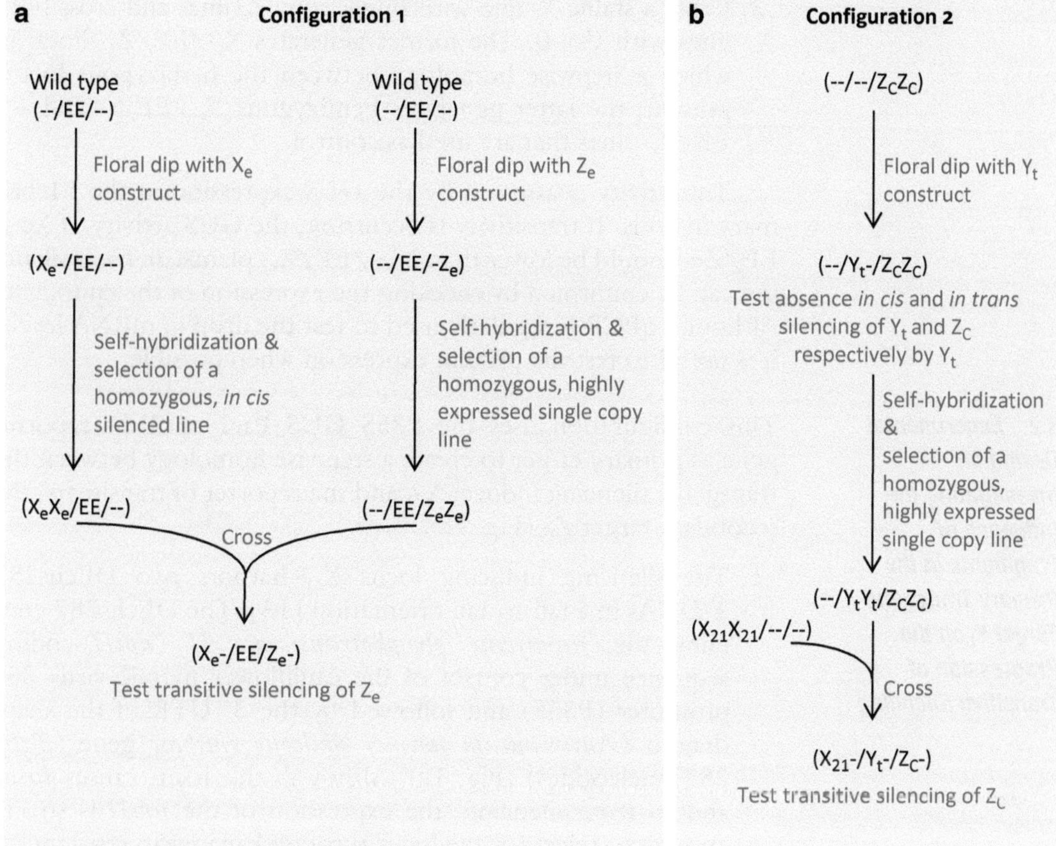


**Fig. 2** Overview of the transgenic lines to be created

or downstream from the region targeted by the silencing inducer to detect 3′–5′ and 5′–3′ transitivity, respectively. As PTGS is investigated, this sequence should be located in the transcribed region. In addition to the P35S_GUS_PacI_npt3'chs reporter gene, the reporter gene T-DNA expresses a chimeric *bialophos acetyltransferase* gene (*bar*), conferring PPT resistance to the plant.

The silencing capacities are controlled by creating combinations of silencing inducer, primary target, and secondary target (Fig. 2, Configuration 1).

1. Create $X_e$ lines specifically silenced for the endogene of interest. Select single locus $X_e$ lines to avoid out-crossing of certain loci and thus variability in the silencing capacity.
2. Transform wild-type Col-0 with the $Z_e$ T-DNA. Select primary transformants with high *GUS* expression and normal expression of the endogene to preclude *in cis* silencing of $Z_e$ leading to *in trans* silencing of the endogene. Select single copy transformants to avoid the establishment of silencing throughout multiple generations.

3. Cross a stable $X_e$ line with single copy $Z_e$ lines and cross both lines with Col-0. The former generates $X_e$-/EE/$Z_e$- lines, in which a stepwise homology between the hemizygous loci is present, the latter generates hemizygous $X_e$-/EE/-- and --/EE/$Z_e$- lines that are used as control.

Transitivity is assessed by the *GUS* expression in the F1 primary hybrids. If transitivity is occurring, the GUS activity of Xe-/EE/Ze- should be lower than in --/EE/$Z_e$- plants. *In trans* silencing can be confirmed by checking the expression of the endogene. Although qPCR is regularly used to test the drop in mRNA levels, it is useful to test the protein expression when possible.

### 3.2 Experimental Design for Investigating the Influence of Fragments in the Primary Transgene Target $Y_t$ on the Progression of Transitive Silencing

This configuration uses the P35S_GUS_PacI_npt3'chs reporter gene as primary target to create a stepwise homology between the transgenic silencing inducer $X_{21}$ and the reporter of transitivity, the secondary target $Z_C$ (Fig. 1c).

1. The silencing inducing locus $X_{21}$ harbors two DEchs287 T-DNAs in a tail-to-tail orientation [13]. The DEchs287 contains the *neomycine phosphotransferase II* (*nptII*)-coding sequence under control of the cauliflower mosaic virus 35S promoter (P35S) and followed by the 3′ UTR of the snapdragon (*Antirrhinum majus*) *chalcone synthase* gene (*3'chs*; 287 nucleotides) (Fig. 1c). Although this locus causes *in cis* and *in trans* silencing, the expression of the *nptII* is strong enough to select for the locus through kanamycin resistance.
2. To construct the primary target $Y_t$, the sequence of interest has to be cloned in the PacI restriction site of the P35S_GUS_PacI_npt3'chs reporter gene. This gene is designed to create a stepwise homology between the silencing inducing locus $X_{21}$ and the secondary target $Z_C$. The *3'chs* UTR preceded by the 3′ part of the *nptII* gene establishes homology between $X_{21}$ and $Y_t$, enabling *in trans* silencing of $Y_t$. The unique PacI restriction site between the *GUS* and *nptII3'chs* sequences allows easy cloning of the region of interest. The *GUS*-coding sequence in this case reports *in trans* silencing. As previously mentioned, the P35S_GUS_PacI_npt3'chs reporter gene also confers PPT resistance.
3. The $Z_C$ locus consists of a single copy of the T-DNA H610, harboring the *GUS*-coding sequence fused to the 3′ UTR of the *nopaline synthase* gene (3'*nos*) under control of P35S, and a chimeric *hygromycin acetyl transferase* (*hpt*) gene conferring resistance to hygromycin [18–20] (*see* **Note 13**). As $Y_t$ and $Z_C$ share the *GUS*-coding sequence, transitive silencing can be detected.

Several combinations of the $Y_t$ locus with $X_{21}$ and $Z_C$ are necessary to investigate the silencing capacities of the system (Fig. 2, Configuration 2).

1. The $Y_t$ T-DNA will be transformed into the homozygous $Z_C$ background. Selection of primary transformants with high *GUS* expression precludes *in cis* silencing of $Y_t$ leading to *in trans* silencing of $Z_C$.
2. From the $Y_t$-/$Z_CZ_C$ transformants, single copy transformants are selected to avoid establishment of *in cis* silencing through the presence of multiple $Y_t$ T-DNA copies throughout multiple generations.
3. Single copy $Y_t$-/$Z_CZ_C$ lines are crossed with the homozygous $X_{21}X_{21}$ line and with Col-0. The former generates $X_{21}$-/$Y_t$-/$Z_C$- lines, in which a stepwise homology between the hemizygous loci is present; the latter generates hemizygous --/$Y_t$-/$Z_C$- lines that are used as controls.
4. Perform a control cross of homozygous $X_{21}X_{21}$ with homozygous $Z_CZ_C$ plants to create hemizygous $X_{21}$-/--/$Z_C$- plants, in which the absence of *in trans* silencing of $Z_C$ directly by $X_{21}$ can be checked.
5. Perform a control cross of homozygous $Z_CZ_C$ plants with Col-0 to create hemizygous --/--/$Z_C$- plants.

To assess transitivity, it is sufficient to measure the *GUS* expression in the F1 primary hybrids. The GUS activity of $X_{21}$-/--/$Z_C$- and --/--/$Z_C$- plants should be similar, as in both lines $Z_C$ should be expressed equally. In hemizygous --/$Y_t$-/$Z_C$- lines, the GUS activity will be slightly higher than in hemizygous --/--/$Z_C$- plants, as the $Y_t$ locus expresses an additional GUS reporter. Only when the GUS activity in $X_{21}$-/$Y_t$-/$Z_C$- hybrids is clearly lower than in hemizygous --/$Y_t$-/$Z_C$- hybrids is transitive silencing across the fragment of interest proven (*see* **Note 14**). When no transitive silencing is recorded, $X_{21}$-/$Y_t$-/-- plants should be checked for *in trans* silencing of the $Y_t$ locus by $X_{21}$. $X_{21}$-/$Y_t$-/-- hybrids can be selected from F2 hybrids produced from self-fertilization of the $X_{21}$-/$Y_t$-/$Z_C$- F1 hybrids by selecting both loci with selective medium and checking the absence of $Z_C$ by PCR (*see* **Note 15**).

### 3.3 A. thaliana Floral Dip Transformation

1. Grow a single colony of the *Agrobacterium* strain containing the appropriate construct on a fresh selective YEB plate for 48 h.
2. In the morning, inoculate 1 mL of $LB^-$ medium without antibiotic in a 50 mL tube with the *Agrobacterium* strain and incubate 8–9 h at 28 °C in a rotary shaker at 230 rpm. Subsequently, add 10 mL of $LB^-$ medium and incubate overnight at 28 °C in a rotary shaker at 230 rpm.
3. The next morning, the OD of the overnight-grown culture is measured at 600 nm in a spectrophotometer. To achieve this, fill a cuvette with 1 mL of a 1/10 dilution of the culture in $LB^-$ medium and mix well before measurement. As a blank, use 1 mL of $LB^-$ medium. The OD should be approximately 2,

corresponding to $10^9$ *Agrobacterium* cells per mL. If the value is too low, the culture should be further incubated at 28 °C and the cell density measured every 30 min.

4. When the required concentration is reached, add 40 mL of freshly made dipping solution (*see* **Note 2**) to the remaining 10 mL of *Agrobacterium* suspension and mix well without vortexing. Use the mix immediately.
5. Dip the flowers of the *Arabidopsis* plants in the suspension and agitate gently for 10 s. On average, the flowers of five plants per *Agrobacterium* suspension are dipped to obtain a sufficient number of independent transformants.
6. Protect the dipped plants from contaminating their neighbors by folding an Aracon tube around them and cover them with Saran wrap for 24 h. Allow plants to grow further in the greenhouse under normal growth conditions: 16 h of light/8 h of darkness, 21 °C.
7. Stop watering the plants ±5 weeks after the floral dip and transfer them to a room at 25 °C with a 20 h light/4 h dark photoperiod and a low humidity. This will decrease the time needed for siliques to dry.
8. When the plants are dry (approximately 8 weeks after floral dip), harvest the T1 seeds (*see* **Note 16**).

### 3.4 Selection and Growth of Transgenic Lines

To select the $Y_t$-/$Z_CZ_C$ and $Z_e$ primary transformants, T1 seeds, harvested from the dipped T0 plants, are sown on medium containing PPT and Hyg, or PPT alone, respectively.

1. Pack ±1,000 seeds (approximately 25 mg) of every dipped *Arabidopsis* plant in miracloth. Keep at −70 °C for 2 days (*see* **Notes 16** and **17**).
2. Surface-sterilize seeds by shaking the miracloth packages for 2 min in 75 % EtOH and upon removal of the EtOH for 7–10 min in 5 % bleach solution containing 0.1 % Tween 20. To obtain a pH of 6–7, the packages are rinsed five times for 5 min in sterile dd$H_2O$, and the pH is checked with pH indicator strips. Before seeding, the packages are soaked in sterile water for about 1 h to enhance germination.
3. Transfer the seeds with a scalpel from a package to a medium plate, pour 2–3 mL 0.1 % agarose over the seeds to spread them. Rotate the plates until the seeds are spread equally over the plate and allow the agarose to solidify before closure with micropore tape.
4. Incubate plates in a growth chamber at 21 °C on a 16 h light/8 h dark cycle.
5. After 14–20 days of growth on the selective medium, only the transformed T1 seedlings will look healthy with two pairs of

green leaves on top of the cotyledons and normally expanded roots. After 3–4 weeks, the selected primary transformants are transferred to soil and grown in the greenhouse (21 °C, 16 h light/8 h dark) until flowering and performing self-fertilization.

6. When siliques are formed, stop watering. When dry, harvest the T2 seeds.

### 3.5 Selection of Single Locus Transgenic Lines

Upon floral dip transformation, several copies of the T-DNA might be inserted, which frequently causes *in cis* silencing of the inserted genes. As this should be excluded for the transgenic primary and secondary targets, lines that highly and stably express the *GUS* reporter gene in a single copy are selected.

1. Harvest leaf material from T1 transformants 6 weeks after Sewing. Determine the *GUS* expression as described in Subheadings 3.8, 3.9, and 3.10 and select the lines with a high *GUS* expression for further analysis (*see* **Note 18**).
2. To identify single locus transformant lines, the segregation ratio in the T2 generation is determined. Of each line selected in Subheading 3.5, **step 1**, 50 seeds of the T2-segregating seed stock are sown on K1 medium with PPT, selective for the corresponding T-DNA selection marker (*see* Subheading 3.4, **steps 2**, **4**, and **5**) (*see* **Notes 19** and **20**). After 3–4 weeks, count the resistant and sensitive seedlings. Single locus lines are expected to have a 3:1 segregation ratio, use a $\chi^2$ statistical test to compare the observed ratio with the expected ratio.
3. From these single locus lines, single copy lines are selected by Southern blot.

### 3.6 Selection of Single Copy Lines Expressing Either $Y_t$ or $Z_e$

When the genomic DNA is digested such that the T-DNA borders are separated into two fragments of unequal length, estimation of the respective fragment length on Southern blot by probing successively with fragment-specific probes will indicate the locus composition. If one copy is present, the length will exceed the expected fragment length because of additional genomic DNA attached. In case of a direct or inverted repeat, the length will exceed the expected fragment length with the fragment of the other T-DNA border or will be duplicated, respectively. The reporter gene T-DNA without insert is split by EcoRV digestion into a fragment containing the LB and the 5′ part of P35S (923 bp), a 5′ *GUS* fragment (952 bp), a middle *GUS* fragment (231 bp) and a fragment containing the 3′ part of the *GUS* gene until the RB (3,913 bp). The suggested probes reside in the P35S and the 3′ part of the *GUS* gene.

#### 3.6.1 Preparation of a DIG Probe by PCR

1. Mix 1 μL non-DIG template, 5 μL 10× PCR buffer, 2.5 μL 50 mM $MgCl_2$, 5 μL DIG dNTPs, 0.25 μL 100 μM forward primer, 0.25 μL 100 μM reverse primer, 0.5 μL 5 U/μL Taq DNA polymerase, and 35.5 μL dd$H_2O$.

2. The PCR reaction consists of 10 min at 95 °C, 30 cycles of 30 s at 95 °C, 45 s at annealing temperature and a variable elongation time at 72 °C and finishes with 7 min at 72 °C and a cool down to 4 °C.
3. Purify the DIG PCR with the High Pure PCR product purification kit, used according to the manufacturer's guidelines. Dissolve in a final volume of 100 μL.
4. Check the probe on an agarose gel: load 5 μL of the purified DIG probe next to some non-DIG probe. The DIG probe will run a little higher than the non-DIG probe due to the incorporated DIG nucleotides.

#### *3.6.2 Blotting the Samples (Wear Gloves at All Times)*

1. Clean all parts of the gel electrophoresis system with 10 % technical SDS, rinse with water, then pour the 1.5 % agarose gel, and use 1× TAE as buffer.
2. Fill the lanes with Smartladder, plasmid DNA or the DNA samples of the selected lines (20–30 μL, 1/10 loading dye). Run the gel for 30 min at 100 V, lower to 25 V, and run overnight. The next morning, the gel is run at 100 V until the colored bands of the loading dye separate the gel into three equal parts.
3. Stain the gel with EtBr for 20 min and take a picture.
4. Clean the glass plates and dishes with 10 % technical SDS and rinse with water.
5. Place the gel in a glass dish and rinse three times with water.
6. Cut off unnecessary parts of the gel and mark the right corner by a cutting.
7. When restriction fragments larger than 10 kb are expected, the DNA is fragmented to facilitate the transfer from gel to membrane. Add 500 mL 0.25 M HCl to the gel in the dish and shake slightly for 10–15 min, until the bromophenol blue marker has turned yellow. Pour off the HCl immediately and rinse with $ddH_2O$.
8. Add 400 mL denaturation buffer and shake for 15 min, then refresh with 600 mL denaturation buffer, and shake again for 15 min. Remove the denaturation buffer and rinse with water.
9. Add 400 mL neutralization buffer and shake for 30 min, refresh with 600 mL neutralization buffer, and shake again for 30 min. Remove the buffer and rinse with water.
10. Clean your working place with Dettol and EtOH. Cut two 15 × 30 cm Whatman papers and a Hybond $N^+$ filter (mark the upper right corner) and two pieces of Whatman paper cut to the size of the gel.
11. Fill a dish with 20× SSC and cover partially with a glass plate. Moisturize the two 15 × 30 cm pieces of Whatman paper in

20× SSC and place on the glass plate with the ends hanging in the 20× SSC solution. Remove air bubbles by rolling with a sterile pipet.

12. Place the gel upside down on the 15 × 30 cm pieces of Whatman paper; subsequently place the Hybond $N^+$ filter on the gel in one fluent movement.
13. Moisturize the two gel-sized pieces of Whatman paper in 20× SSC, place them on the filter, and remove air bubbles by rolling with a pipet.
14. Surround the gel and the 15 × 30 cm pieces of Whatman paper with double-folded Saran Wrap to prevent the napkins from touching either them or the 20× SSC.
15. Place the napkins (folded over four times to equalize the height), the second glass plate, and a weight of about 500 g on the gel-sized pieces of Whatman paper.
16. Incubate overnight (±16 h) and take a picture of the gel to ensure that all DNA is transferred.
17. Dry the filter for ±1 h on Whatman paper, DNA side facing upwards.
18. Fix the DNA on the filter by UV cross-linking in the Biorad GS genelinker UV chamber, using program C2 (DNA side facing upwards).

#### 3.6.3 Hybridization and Detection

1. Wash the hybridization tubes with 10 % SDS and rinse with $ddH_2O$.
2. Place the filter in a hybridization tube with the cross-linked DNA facing the inside and add 15–40 mL DIG Easy Hyb buffer. Prehybridize for 1 h at 42 °C.
3. In the meantime, denature the 100 μL DIG probe for 5 min at 95 °C, transfer onto ice immediately to prevent renaturation and keep it on ice for at least a few minutes (*see* **Notes 21** and **22**).
4. Add the denatured probe to the DIG Easy Hyb buffer, but not onto directly on the filter! Hybridize overnight at 42 °C (*see* **Note 22**).
5. Pour off the hybridization solution (*see* **Note 22**).
6. Wash the filter 2× 5 min in 100 mL 2× SSC/0.1 % SDS at room temperature.
7. Wash the filter 1× 15 min in 2× SSC/0.1 % SDS at 65 °C (preheat this buffer).
8. Wash the filter 1× 15 min in 1× SSC/0.1 % SDS at 65 °C (preheat this buffer).
9. Take the filter out of the hybridization tube and put it in a Petri dish.

10. Wash 1× 5 min in wash buffer at room temperature on a shaker.
11. Pour of the wash buffer and cover the filter with block buffer. Incubate on the shaker for 30 min.
12. In the meantime, prepare the block anti-DIG solution: centrifuge the anti-DIG-AP and add 2 μL of the antibody to 40 mL block buffer.
13. Replace the block buffer with the block anti-DIG solution and incubate on the shaker for another 30 min.
14. Pour of the block anti-DIG solution and wash the filter at least 3× 15 min in wash buffer.
15. Incubate the filter for 5 min in B3 buffer.
16. Replace the B3 buffer with B3 buffer containing CDP-Star. Clean the bench with EtOH and Dettol. Cut open a freezer bag along two sides and attach half of the bag to the bench with tape to obtain a book-like structure. Use pincers to place the filter on the attached side with the DNA facing up. Pour the CDP-Star solution (60 μL CDP-Star/6 mL B3 buffer) on the bottom of the blot. Close the bag softly to help the solution to spread across the membrane. Incubate 5 min at room temperature. Remove the solution by wiping the solution with a tissue to a corner of the bag.
17. Remove the filter from the bag, dip-off on Whatman paper for a few seconds, put the filter in a second cut-open freezer bag taped in a cassette, seal the plastic, and expose to a film for a few minutes (5–30 min) in a dark room.

#### *3.6.4 Strip the Membrane and Reprobe*

1. Rinse the membrane for 1 min in $ddH_2O$.
2. Wash twice with 0.2 M NaOH/0.1 % SDS for 15 min at 37 °C.
3. Rinse twice in 2× SSC for 5 min.
4. Store semi-wet in a sealed plastic bag at 4 °C or reprobe immediately by repeating Subheading 3.6.3.

### 3.7 Crossing of Lines Expressing the Silencing Inducer with Lines Expressing the Primary and Secondary Target

To complete the XYZ configuration, plants containing the silencing inducing locus $X_{21}X_{21}$ or $X_eX_e$/EE have to be crossed with the selected $Y_tY_t/Z_CZ_C$ or EE/$Z_eZ_e$ lines, respectively (*see* Fig. 2 and **Note 23**).

1. Grow parental plants on selective medium as described in Subheading 3.5, **step 2**, but use less seed.
2. Transfer five male and 15 female parental plants to soil 2–3 weeks after germination and grow them up to 7 weeks in greenhouse conditions. Protect them from cross-fertilization with Aracon bases and tubes.
3. Select buds of the female parental plant where the tip of the stigma is just poking out, the anthers should be immature. Emasculate these buds by removing the anthers.

4. Remove a mature anther from the male parental plant from a flower that is just opening. Rub this anther over the stigma of the female flower (*see* **Note 24**).
5. Label the pollinated female flowers with Sewing thread. If crossing is successful, an elongation of the stigma is obvious after 2 days.
6. Two weeks after crossing, cut off the siliques and transfer to an Eppendorf with a little hole in the cap and dry them at 28 °C.

The F1 hybrids are selected by Sewing on medium selective for all loci expected to be present. Resistant plants are transferred to soil 3 weeks after Sewing and the GUS activity is generally measured 6 weeks after Sewing.

### 3.8 Protein Extraction from Arabidopsis Leaves

1. Harvest two to three healthy green rosette leaves of 6-week-old plants grown in soil. Collect the material in a 2 mL Eppendorf tube containing two metal balls and immediately store in liquid nitrogen to prevent degradation.
2. Cool down the grinding ball mill adapters in liquid nitrogen, insert the samples, and shake 30 s at 30 Hz.
3. After grinding, keep the samples on ice during the rest of the protocol to prevent breakdown of the proteins.
4. Short-spin centrifuge at 4 °C, 16,000×*g* to spin down the leaf material.
5. Add 200 μL GUS EB 2×, mix well, and incubate on ice for 5–10 min.
6. Centrifuge 10 min at 4 °C, 16,000×*g* and collect the supernatant in a new 1.5 mL Eppendorf.
7. Centrifuge 10 min at 4 °C, 16,000×*g* and collect 180 μL supernatant in a new 1.5 mL Eppendorf.
8. Add 45 μL (1/4 of the collected supernatant volume) glycerol and pipet up and down to homogenize (*see* **Note 25**).
9. Store at −70 °C.

### 3.9 Determination of the Protein Concentration in Leaf Extracts

The solubilized protein concentration is determined with the Bio-Rad Protein Assay, based on the method of Bradford [21]. Comparison to a standard curve made of a BSA dilution series provides a relative measurement of the protein concentration.

1. To determine the protein concentration in the samples, use a BSA dilution series as standard. Fill the first two columns of the plate with:
   (a) 25 μL GUS EB 2× 1/5 (blank)
   (b) 1 μg/mL: 2.5 μL 0.1 mg/mL BSA + 17.5 μL $ddH_2O$ + 5 μL GUS EB 2×

(c) 2 μg/mL: 5 μL 0.1 mg/mL BSA + 15 μL $ddH_2O$ + 5 μL GUS EB 2×

(d) 4 μg/mL: 10 μL 0.1 mg/mL BSA + 10 μL $ddH_2O$ + 5 μL GUS EB 2×

(e) 6 μg/mL: 15 μL 0.1 mg/mL BSA + 5 μL $ddH_2O$ + 5 μL GUS EB 2×

(f) 8 μg/mL: 20 μL 0.1 mg/mL BSA + 5 μL GUS EB 2×

(g) 10 μg/mL: 25 μL 0.1 mg/mL BSA + 5 μL GUS EB 2×

2. Prepare a 1/10 dilution of each protein extract in sterile $ddH_2O$. Bring 5 μL and 10 μL in neighboring wells, add 20 μL and 15 μL GUS EB 2× 1/5 (final dilution of 500× and 250×), respectively.
3. Add 225 μL BIO-RAD reagents to each well. Seal the plate with a sticker and shake at 400 rpm for 2 min at room temperature.
4. Read the absorbance in a spectrophotometer at 600 nm, preceded by a 5 s shake. With the Softmax Pro software, calculate the protein concentrations from the slope of the BSA standard dilution series.

### 3.10 Fluorimetric Assay of the GUS Activity in Leaf Extracts

GUS hydrolyzes the non-fluorescent substrate 4-methylumbelliferyl-β-D-glucuronide hydrate (4-MUG) into equimolar amounts of the fluorescent product 4-MU. By measurement of the fluorescence upon addition of an excess amount of the substrate and comparison with a standard curve of GUS enzyme, the GUS activity in the protein extracts can be determined [22]. This activity is expressed in Units per mg total soluble protein (U/mg TSP), where 1 U GUS is the amount needed to produce 1 nmol product at 37 °C in 1 min.

1. The GUS activity of the samples should fall within the linear range of the standard curve; therefore, the protein extracts are diluted in GUS EB 2×. Make four dilutions for each protein extract, depending on the expected GUS activity.

(a) When a relatively low GUS activity is expected: 250× = 4 μg/10 μL; 500× = 2 μg/10 μL; 1,000× = 1 μg/10 μL; 4,000× = 0.25 μg/10 μL

(b) When an intermediate GUS activity is expected: 500× = 2 μg/10 μL; 2,000× = 0.5 μg/10 μL; 5,000× = 0.2 μg/10 μL; 20,000× = 0.05 μg/10 μL

(c) When a high GUS activity is expected: 5,000× = 0.2 μg/10 μL; 10,000× = 0.1 μg/10 μL; 20,000× = 0.05 μg/10 μL; 100,000× = 0.01 μg/10 μL

2. Make the dilution series used for the standard curve:
   (a) First make a 3.6 U/μL GUS enzyme stock solution. Add 1 mL glycerol to 1 mL GUS EB 2×, add 190 μL of this solution to 10 μL GUS enzyme 72 U/μL, and store at −20 °C.
   (b) Generate the following dilutions from the stock solution made in **step a**:
      (i) 100 mU/μL: 5 μL 3.6 U/μL + 175 μL GUS EB 2×
      (ii) 10 mU/μL: 10 μL 100 mU/μL + 90 μL GUS EB 2×
      (iii) 1 mU/μL: 10 μL 10 mU/μL + 90 μL GUS EB 2×
      (iv) 0.1 mU/μL: 10 μL 1 mU/μL + 90 μL GUS EB 2×
3. Load 10 μL of the protein extract dilutions into individual wells.
4. Load the standard curve in duplicate:
   (a) Blanco: 10 μL GUS EB 2×
   (b) 2 mU GUS: 20 μL 0.1 mU/μL
   (c) 5 mU GUS: 5 μL 1 mU/μL
   (d) 10 mU GUS: 10 μL 1 mU/μL
   (e) 20 mU GUS: 20 μL 1 mU/μL
   (f) 50 mU GUS: 5 μL 10 mU/μL
   (g) 100 mU GUS: 10 μL 10 mU/μL
5. As a detection control, fill one well with 12 μL 1 mM 4-MU (12 nmol 4-MU) and one with 24 μL 1 mM 4-MU (24 nmol 4-MU).
6. Add 240 μL GUS EB 2× to all wells and 10 μL 4 mM 4-MUG to all wells except those filled with 4-MU.
7. Seal the plate with a microtiter plate sticker, protect from light with aluminum foil, and shake thoroughly on a rotational shaker for 3 min.
8. Remove the sticker and aluminum foil and incubate 10 min in the preheated (37 °C) Fluostar OPTIMA (*see* **Note 26**).
9. Measure the fluorescence intensity with the FLUOstar OPTIMA. Use the excitation filter 355-ex and the emission filter 450-10. Before reading 16 cycles at 37 °C, the plate should be shaken for 10 min at 600 rounds per min.
10. Check in the signal curve window if the measurements of cycles two to six fall within the linear part and export the slope/min in raw data—blank format. Use the standard curve to calculate the amount of mU GUS/well. Taking into account the dilution factor, the amount of mU GUS/mg TSP can be calculated (*see* **Note 27**).

## 4 Notes

1. To grow *Arabidopsis* plants for floral dip transformation, seeds are sown in regular soil and plants are grown at 22 °C (day) and 18 °C (night) with a photoperiod of 12 h of light/12 h of darkness. To obtain more floral buds per dipped plant, the inflorescences are cut off when most plants have formed primary bolts. This removal will trigger the plant to produce synchronized multiple secondary bolts. When dipped, the plants should have several secondary flowering stems of about 10 cm and immature flowers and flower buds (approximately 2 weeks after removing the primary inflorescence).
2. The surfactant Silwet should be mixed thoroughly into the 10 % sucrose solution. The mix cannot be stored and should be used immediately for the floral dip.
3. Murashige and Skoog modified Vitamin Mix (Duchefa) can also be added to the growth medium, but is not necessary. Add 1 mL 1,000× vitamin stock (10.4 g vitamins in 100 mL water (*see* **Note 4**)) to 1 L of K1 medium.
4. The stock solutions of selective agents are filter sterilized and stored as aliquots in a −20 °C freezer. As the kanamycin solution is sensitive to freeze-thaw cycles, defrosted aliquots cannot be frozen again.
5. Instead of carbenicilline, 800 mg vancomycin can be added as such to the 60 °C sterilized K1 medium.
6. Tween 20 is a detergent used to lower the surface tension and as such enhance the washing.
7. Use a proofreading PCR, for instance, $Vent_R$ DNA Polymerase of New England BioLabs (Ipswich, MA).
8. We used the EcoRV of Promega (Madison, WI) in the supplied buffer D. Digest 5 μg genomic DNA with 15 U EcoRV (1.5 μL of 10 U/μL) in 1× buffer D (2 μL 10× buffer D) supplemented with 2 μg BSA (0.2 μL 10 μg/μL) in a total volume of 20 μL (add $H_2O$ up to this volume).
9. An equal amount of plasmid copies compared to genomic DNA copies is loaded on the gel. As the *A. thaliana* genome is about $1.25 \times 10^8$ bp and the plasmid without insert is 12,153 bp, 0.097 ng should be loaded. A double quantity is used to ensure a strong enough signal.
10. The phosphate buffer is made freshly on the day of use. The 0.2 M $NaH_2PO_4$ and 0.2 M $Na_2HPO_4 \cdot 12H_2O$ can be stored for upon 1 year at room temperature when filter sterilized and kept sterile.
11. The 4-MU solution is light sensitive. It should be made freshly and protected from light with aluminum foil.

12. Can be stored for longer periods at −20 °C.
13. The H610 T-DNA was previously described as the H2 or H T-DNA [18, 19]. The transgenic plant line homozygous for the single T-DNA copy $Z_C$ locus (FH33/1) was described in [20].
14. When transitivity is strong, the *GUS* expression will be below the detection limit. However, intermediate transitivity with inter-planta variability ranging from low to high activity can occur and transitivity also accumulates over time. When dealing with intermediate transitivity, we advise to measure the GUS also some weeks later. In general we measure GUS activity 6 weeks after germination.
15. Alternatively, the presence of $Z_C$ can be checked by a callus induction test on hygromycin-containing (20 mg/L) medium.
16. Seed stocks can be stored for a long time at room temperature or at 4 °C, but they should be kept dry. When the seeds are stored at room temperature, vernalization by either storing the seeds for 24 h at −70 °C before Sewing (introduced into the protocol to reduce the egg survival of thrips within the seeds) or storing the sown plates 1–4 days at 4 °C is recommended to stimulate germination.
17. In our department, the transformation efficiency of the floral dip transformation varies between 0.5 and 2 %, indicating that Sewing approximately 1,000 seeds results in 5 20 transformants.
18. With our constructs, the *GUS* expression in the T1 transformants varies between below the detection limit and about 150–500 U GUS/mg TSP, depending on the construct used and the line it was dipped in. Bear in mind that dipping in the $Z_CZ_C$ background that already expresses two *GUS* genes leads to higher maximum values. The presence of a super-transformed $Y_t$ locus that is silenced will however cause *in trans* silencing of the $Z_C$ locus, which will result in a clearly lower expression than in the control $Z_CZ_C$ line. We select the top 15–20 lines having the highest expression to continue.
19. According to our protocol, over 70 % of the T1 transformants obtained through floral dip transformation contain one or more T-DNAs integrated at one genetic locus [23].
20. To facilitate the count of resistant plants, a regular spacing of seeds allowing growth without overlap is advised. To achieve this, plates with a grid on the bottom are convenient, as one can place one seed per square or intersection with a pair of tweezers or a sterile toothpick.
21. ±50 μL probe/membrane can be used when hybridizing multiple filters.
22. To reuse a probe that is already in DIG Easy Hyb buffer, collect in a 50 mL tube after hybridization and store at −20 °C.

Before use, denature the probe in a 68 °C water bath for 10 min and put on ice for 10 min. Upon prehybridization, remove the DIG Easy Hyb buffer and replace by the probe denatured in DIG Easy Hyb buffer. The probe can be reused three to four times.

23. In general we cross two single copy $Y_tZ_C$ lines with the $X_{21}$ line. As such, the influence of position effects is minimized.
24. When using a suitable anther, the pollen should visibly stick to the stigma.
25. Glycerol is viscous and therefore difficult to pipet. We cut off the lower 2 mm of the 200 μL tips to enhance pipetting and mixture.
26. By incubation at 37 °C, the GUS enzyme can work at its optimal temperature.
27. Measurements that are out of the linear range of the standard curve, more specifically those that are below or above the measurements of the lowest and highest standard dilution, respectively, are not taken into account.

## Acknowledgments

The authors thank Annick Bleys for help in preparing the manuscript. Many thanks to Sylvie De Buck for advice. This work was supported by a grant from the Research Foundation-Flanders (no. G.0211.06N). LV was indebted to the Agency for Innovation through Science and Technology (IWT-Vlaanderen) for a predoctoral fellowship.

### References

1. Cerutti H, Casas-Mollano JA (2006) On the origin and functions of RNA-mediated silencing: from protists to man. Curr Genet 50: 81–99
2. Bleys A, Van Houdt H, Depicker A (2006) Transitive and systemic RNA silencing: both involving an RNA amplification mechanism? In: Nellen W, Hammann C (eds) Small RNAs: analysis and regulatory functions. Springer, Berlin
3. Himber C, Dunoyer P, Moissiard G, Ritzenthaler C, Voinnet O (2003) Transitivity-dependent and -independent cell-to-cell movement of RNA silencing. EMBO J 22: 4523–4533
4. Nicolás FE, Torres-Martínez S, Ruiz-Vázquez RM (2003) Two classes of small antisense RNAs in fungal RNA silencing triggered by non-integrative transgenes. EMBO J 22:3983–3991
5. Sijen T, Fleenor J, Simmer F, Thijssen KL, Parrish S, Timmons L, Plasterk RHA, Fire A (2001) On the role of RNA amplification in dsRNA-triggered gene silencing. Cell 107: 465–476
6. Van Houdt H, Bleys A, Depicker A (2003) RNA target sequences promote spreading of RNA silencing. Plant Physiol 131:245–253
7. Voinnet O (2008) Use, tolerance and avoidance of amplified RNA silencing by plants. Trends Plant Sci 13:317–328
8. Braunstein TH, Moury B, Johannessen M, Albrechtsen M (2002) Specific degradation of 3′ regions of GUS mRNA in posttranscriptionally silenced tobacco lines may be related to 5′–3′ spreading of silencing. RNA 8:1034–1044
9. Petersen BO, Albrechtsen M (2005) Evidence implying only unprimed RdRP activity during

transitive gene silencing in plants. Plant Mol Biol 58:575–583

10. Deleris A, Gallego-Bartolome J, Bao J, Kasschau KD, Carrington JC, Voinnet O (2006) Hierarchical action and inhibition of plant Dicer-like proteins in antiviral defense. Science 313:68–71
11. Miki D, Itoh R, Shimamoto K (2005) RNA silencing of single and multiple members in a gene family of rice. Plant Physiol 138: 1903–1913
12. Vaistij FE, Jones L, Baulcombe DC (2002) Spreading of RNA targeting and DNA methylation in RNA silencing requires transcription of the target gene and a putative RNA-dependent RNA polymerase. Plant Cell 14:857–867
13. Bleys A, Van Houdt H, Depicker A (2006) Down-regulation of endogenes mediated by a transitive silencing signal. RNA 12:1633–1639
14. Van Houdt H, Ingelbrecht I, Van Montagu M, Depicker A (1997) Post-transcriptional silencing of a neomycin phosphotransferase II transgene correlates with the accumulation of unproductive RNAs and with increased cytosine methylation of 3′ flanking regions. Plant J 12:379–392
15. Bleys A, Vermeersch L, Van Houdt H, Depicker A (2006) The frequency and efficiency of endogene suppression by transitive silencing signals is influenced by the length of sequence homology. Plant Physiol 142:788–796
16. Vermeersch L, De Winne N, Depicker A (2010) Introns reduce transitivity proportionally to their length, suggesting that silencing spreads along the pre-mRNA. Plant J 64:392–401
17. Bleys A, Karimi M, Hilson P (2009) Clone-based functional genomics. In: Belostotsky DA (ed) Plant systems biology, vol 553, Methods in molecular biology. Humana Press, New York
18. De Buck S, Jacobs A, Van Montagu M, Depicker A (1998) *Agrobacterium tumefaciens* transformation and cotransformation frequencies of *Arabidopsis thaliana* root explants and tobacco protoplasts. Mol Plant Microbe Interact 11:449–457
19. De Buck S, Van Montagu M, Depicker A (2001) Transgene silencing of invertedly repeated transgenes is released upon deletion of one of the transgenes involved. Plant Mol Biol 46:433–445
20. De Buck S, Windels P, De Loose M, Depicker A (2004) Single-copy T-DNAs integrated at different positions in the *Arabidopsis* genome display uniform and comparable β-glucuronidase accumulation levels. Cell Mol Life Sci 61: 2632–2645
21. Bradford MM (1976) A rapid and sensitive method for the quantitation of microgram quantities of protein utilizing the principle of protein-dye binding. Anal Biochem 72:248–254
22. Breyne P, De Loose M, Van Dedonder A, Montagu M, Depicker A (1993) Quantitative kinetic analysis of β-glucuronidase activities using a computer-directed microtiter plate reader. Plant Mol Biol Rep 11:21–31
23. De Buck S, Podevin N, Nolf J, Jacobs A, Depicker A (2009) The T-DNA integration pattern in Arabidopsis transformants is highly determined by the transformed target cell. Plant J 60:134–145

# Index



Charles Spillane and Peter C. McKeown (eds.), *Plant Epigenetics and Epigenomics: Methods and Protocols*, Methods in Molecular Biology, vol. 1112, DOI 10.1007/978-1-62703-773-0, © Springer Science+Business Media New York 2014

## Q



## R



## S



## T



## U



## V



## W



## Z

MIX
Papier aus verantwortungsvollen Quellen
Paper from responsible sources
FSC® C105338

If you have any concerns about our products,
you can contact us on
**ProductSafety@springernature.com**

In case Publisher is established outside the EU,
the EU authorized representative is:
**Springer Nature Customer Service Center GmbH**
**Europaplatz 3, 69115 Heidelberg, Germany**

Printed by Libri Plureos GmbH
in Hamburg, Germany